D0706197

Build Your Own Fuel Cells

by Phillip Hurley

copyright ©2002, 2013 Phillip Hurley
all rights reserved

illustrations and photography
copyright ©2002, 2013 Good Idea Creative Services
all rights reserved

ISBN-13: 978-0-9837847-6-0

Wheelock Mountain Publications
is an imprint of
Good Idea Creative Services
Wheelock VT
USA

Copyright ©2002 by Phillip Hurley and Good Idea Creative Services

First print edition 2013

Notice of Rights

All rights reserved. No part of this book may be reproduced or transmitted in any form or by any means, electronic, mechanical, photocopying, recording or otherwise without prior written permission of the publisher. To request permission to use any parts of this book, please contact Good Idea Creative Services, permission@goodideacreative.com.

Wheelock Mountain Publications is an imprint of:

Good Idea Creative Services
324 Minister Hill Road
Wheelock VT 05851 USA

ISBN   978-0-9837847-6-0

Library of Congress Control Number: 2013937354

Library of Congress subject headings:

Fuel cells--Design and construction
Fuel cells--Handbooks, manuals, etc.
Fuel cells--Amateur's manuals

Disclaimer and Warning

The reader of this book assumes complete personal responsibility for the use or misuse of the information contained in this book. The information in this book may not conform to the reader's local safety standards. It is the reader's responsibility to adjust this material to conform to all applicable safety standards after conferring with knowledgeable experts in regard to the application of any of the material given in this book. The publisher and author assume no liability for the use of the material in this book as it is for informational purposes only.

Contents

Contents

Contents

Experimental Methods for Making MEAs

Fuel Cell Basics

A Short History of the Fuel Cell

The fuel cell is over one hundred and seventy years old.

In 1839 William Grove was experimenting with batteries and the electrolytic process when he came upon the idea of reversing the process to generate electricity. Electrolysis uses electricity to separate hydrogen atoms from oxygen atoms in water molecules. Grove's idea was to instead combine oxygen and hydrogen to produce electricity, and water. With the use of a platinum catalyst electrode for this purpose, his efforts were successful. This research was fundamental to understanding the principals of the operation of a fuel cell.

Grove's "gas battery"

Grove called his invention a "gas battery." His fuel cell consisted of two platinum electrodes immersed in an electrolyte solution of sulfuric acid and water. This is the same type of electrolyte that is used in lead acid rechargeable batteries today.

Grove was also the first person to put the fuel cell to practical work. In 1842 he connected fifty of his fuel cells together to power an electric arc.

The gas battery did not attract much attention even though Grove proved that it was a practical means of generating electricity. Many years later, in 1889, Ludwig Mond and Charles Langer began experimenting with Grove's original concept. They were responsible for popularizing the term "fuel cell" to refer to the Grove gas battery. The name stuck and that is what we call them today. Mond and Langer's fuel cells used platinum electrodes with a clay barrier soaked in a sulfuric acid electrolyte.

During this time Frederick Ostwald also researched and developed theories on the operation of the fuel cell, and later, during the first half of the 20th century, Emil Baur did experimental work with a variety of fuel cell configurations, using different electrolytes such as molten silver and clay mixed with metal oxides. His research was used by the British government to develop fuel cells for submarines.

In the late 1930s Francis Bacon designed and experimented with a fuel cell that had nickel gauze electrodes. Much of his later research and work found its way into the Apollo space program. In the 1950s Harry Karl Ihrig built the first fuel cell powered tractor.

Fuel cells today

During the 1980s and '90s fuel cell development took off in a big way and many small fuel cell power plants were built. Today fuel cells are used in the NASA Space Shuttle as well as in terrestrial stationary and mobile applications with many new developments underway. They are more efficient and cleaner-running than the internal combustion engine and in the near future will probably replace those engines in motor vehicles. Fuel cells are now powering everything from laptop computers, cars and buses to home and industrial electrical systems.

What Exactly Is a Fuel Cell?

Simply put, a fuel cell is an energy conversion device. There are no moving parts involved; thus fuel cells operate silently. The energy released as heat and electricity can be used as a power source.

The process begins with feeding hydrogen to one catalyst electrode which facilitates the separation of the hydrogen atoms into electrons and protons. The protons or hydrogen ions move through the membrane towards the other catalyst, which is being fed with oxygen. The stripped electrons cannot pass through the membrane or electrolyte, so they must be routed through an external circuit. The external circuit contains an electrical load such as a motor or light bulb, etc., and leads to the other catalytic electrode, where the protons and electrons recombine and bond with oxygen to create water molecules.

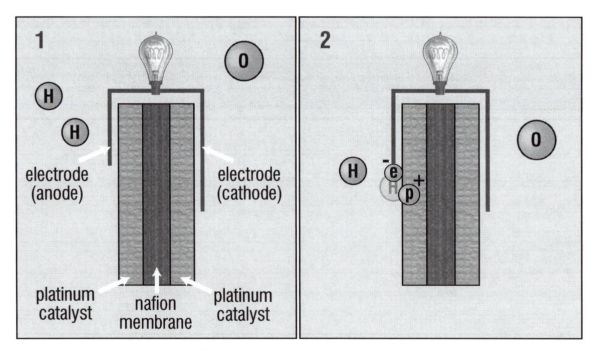

How a PEM fuel cell works

Each hydrogen atom separates into an electron and a proton at the catalyst layer.

What Exactly is a Fuel Cell?

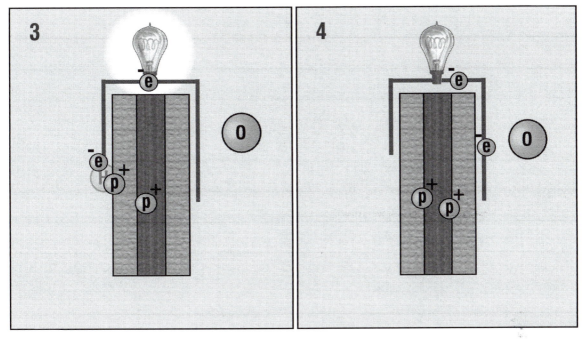

The electrons are routed through the anode to an electrical circuit with an electrical load

The protons move through the membrane towards the cathode

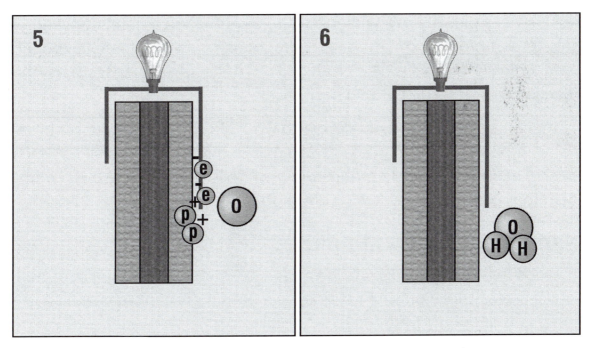

At the cathode, the electrons and protons recombine into hydrogen atoms and bond with oxygen to form water.

Fuels for Fuel Cells

Many sources of hydrogen

All fuel cells use hydrogen as a fuel. Hydrogen is the most abundant element in the universe, and fuel cells are quite versatile as to what forms of fuel they can use. On our planet hydrogen is not a free floater but is combined with other elements and locked in compounds. So, there are many different sources of hydrogen, and the method of liberating the hydrogen for fuel cell consumption can be "green" and clean, or still involve polluting by-products.

Fuel delivery packages

Some examples of fuels that hydrogen can be extracted from in order to power fuel cells ("fuel delivery packages") are: water, methane, ethane, toluene, butene, butane methanol, gasoline, propane and natural gas. Various plants such as algae when oxygen starved and deprived of sulphur will produce an enzyme, hydrogenase, which extracts hydrogen from water. Hydrogen is also produced by chemical reactions such as the reaction between sulfuric acid and zinc.

There are DMFC type fuel cells which work directly off methanol or ethanol with no reforming process necessary, and there are fuel cells which work directly off sodium borohydrate. Fuel cells can also get their hydrogen from metal hydrides, which are a good hydrogen storage medium.

Hydrogen from fossil fuels

When fossil fuels are used to power fuel cells, the fuel must be reformed, which creates pollutants; but the noxious emissions from reformation are far less than what is produced by internal combustion engines burning fossil fuels. Though they are a good source of hydrogen, fossil fuels are limited in quantity, and not a renewable energy source. They are hazardous to transport and store, and are toxic.

Hydrogen from water

Water has none of these bad habits, which makes it the most desirable fuel package available. If water is the fuel package, no pollutants are created. Water, when split into hydrogen and oxygen and then recombined in a fuel cell, produces water – infinitely renewable. Water covers about 70% of the earth's surface and is perhaps the most abundant hydrogen package on the planet. It does not require transportation over thousands of miles as fossil fuels often do. Water does not pollute if spilled – it is environmentally benign and one of the safest storage mediums as it does not burn. This is why water will be the key to the future hydrogen economy.

Technology of the future

Because of their versatility, fuel cells will soon be everywhere in our daily lives. The transition to a hydrogen-based economy can be made fairly easily because the infrastructure of our fossil-fuel based economy can accommodate fuel cells that run on gasoline or methanol. The fossil fuel infrastructure has funded much research in this field, and this technology that was largely ignored for more than a hundred years is now developing quickly. But, in time we will transition to a hydrogen economy totally based on oxygen-hydrogen fuel cells that are truly non-polluting.

Global impact

Fuel cells and the hydrogen economy will have a huge impact on "developing" nations and the global balance of power. Hopefully where "modern infrastructures" are lacking, some nations will be able to skip the oil addiction phase and "develop" directly into non-polluting, decentralized hydrogen economies. Perhaps there will be fewer wars altogether as we all grow out of our dependence on oil from far-flung and politically unstable regions of the world – hydrogen can be made at home, or right at the gas station!

Storing hydrogen

Pure hydrogen can be stored as a liquid, gas, or in metal hydrides which can be titanium, iron, manganese or other metal compounds.

Hydrogen can also be produced on demand, which cannot be said for fossil fuels.

Hydrogen can be stored as safely as any other gas or fuel, but it requires more storage space as it does not compress to favorable energy per volume as do other gases, such as propane. This is not a problem for stationary installations such as home power units, etc., but for vehicles, metal hydrides will take the forefront. The possibilities for the future are limitless. There is much research to be done yet in the quest for more efficient and economical release of hydrogen from water.

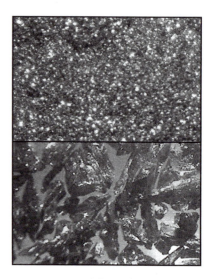

Metal hydrides, above, and hydride storage bottle for hydrogen, below

What is a fuel?

A fuel is defined by current general consensus as any material such as coal, oil, gas, wood, etc., that is burned to supply heat or power. However, it's obvious that this definition is not adequate to describe the operation of a fuel cell, because hydrogen is not burned in a fuel cell. (As an aside, the term "fuel cell" could be considered a misnomer. Groves' term, "gas battery" is actually more accurate.) Hydrogen is considered a fuel because it can be burned in the presence of oxygen.

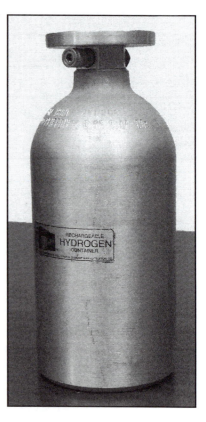

In fact a fuel is not a fuel until it is combined with oxygen – that is, it cannot be "burned" unless oxygen is present – so, perhaps it would be more correct to say that a fuel is a substance or compound combined with oxygen. Or, perhaps it would be more correct to call fuel cells "gas batteries," their original name, per Groves.

Water is a fuel

Anyway, a fuel cell can use a variety of substances usually considered fuels, or use water as a fuel. Classically speaking, water is not considered a fuel because it does not burn. But, for fuel cells, a fuel is any substance from which hydrogen can be extracted, so, as far as fuel cells are concerned, water is fuel. As a matter of fact, water comes with its own oxidizer – oxygen – and is perhaps one of the neatest and safest storage packages available.

Hidden inefficiencies of fossil fuels

The established fossil fuel industry and its proponents have a very hard time acknowledging that water is a fuel which is not only safe to store and handle but also readily available in most locations. They don't like to talk about water, and will tell you that the methods used to split water are inefficient and expensive when compared with extraction of hydrogen from fossil fuels. Electrolysis units are in reality about 75% efficient. As far as fossil fuels are concerned, natural gas has the best hydrogen to carbon ratio. The hydrogen can be extracted by steam reforming with efficiencies of 70-90%. What does not show up in the statistics offered by the fossil fuel establishment is that fossil fuel reforming produces pollution.

Who bears the real costs?

The costs of storage, shipping and piping are very high for fossil fuels. Then there are the regulators in local, state and federal agencies required to enforce safe standards for storing, shipping or piping these fuels. Not to mention the subsidies to develop oil and gas fields that come out of the taxpayers' pockets, and the privately funded development money for such exploration which is sure to eventually come from the consumer's pockets.

It also costs a lot of money to clean up the inevitable fossil-fuel leaks and spills – if indeed they can ever be cleaned up adequately. The damage to the ecosystem does not show up in the ledgers and reports of the fossil fuel industry. The real costs are unfortunately generally not the burden of those who gain financially by the processes which create the pollution. The real costs are born by society and the planet at large.

All the previous mentioned issues dissolve when the fuel is water. Think about it!

I've barely touched on the negatives of the fossil fuel economy. Suffice it to say that the industry's manipulation of statistics of costs do not take into account the many and very big hidden costs associated with the continued use of fossil fuels.

Skeletons in the closet

One last note – the wars fought and the lives lost, and the cost of such activities (deemed necessary to protect our oil supplies) is also never figured into the fossil fuel industries statistics. These hidden costs can aptly be called skeletons in the closet of the fossil fuel economy.

The hydrogen-water economy is coming!

There really is no comparison. Water is a much cheaper alternative to fossil fuels. Using fossil fuels for fuel cells is only a stop-gap measure – the limited supply of fossil fuels will end the fossil fuel economy sooner than later. It behooves us to move along and quickly accept water as a "fuel" – we really don't have a choice. It is certain that a hydrogen-water economy is on the horizon, as it is infinitely renewable and pollution free. So, let's get on with it!

Water based hydrogen production

Fossil fuel based hydrogen production depends on the steam reforming of natural gas to produce hydrogen economically. Water-based hydrogen production depends on electrolysis and variants such as photo electrolysis, and photo-biological techniques. Hydrogen can also be extracted from biological matter through a gasification and steam reformation process; that is to say, you can also make hydrogen from garbage and compost.

Since this book is not particularly concerned with the reformation of fossil fuels, or biomass gasification and hydrogen extraction, we will focus on other methods.

Electrolyzers

Electrolysis is the proven method for the production of hydrogen from water. There are basically two types of electrolyzers, tank electrolyzers and filter press electrolyzers. Tank electrolyzers are very simple to construct and the tolerances are not critical. Filter press electrolyzers are more expensive to build but they are more efficient and take up less space. Both types are good for their own reasons and both have a place in the scheme of things.

Solar, wind and hydro power for electrolyzers

Electrolyzers need a DC power source. Photovoltaic, wind or hydro power systems are all excellent providers of DC for electrolyzers.

Photo-electrolysis

There is a variant of photovoltaic power supply systems which is called photo-electrolysis. Photo cells are immersed in an electrolyte solution or water, and exposed to light to generate hydrogen. This technique is highly experimental and not presently very efficient, but progress is being made. Photo catalysts for this process usually consist of mixtures of the metals titanium, niobium or tantalum which are able to generate hydrogen and oxygen in water when exposed to sunlight. Some other mixtures used in this process consist of rubidium, neodymium, tantalum and oxygen.

Photo-biological

For photo-biological hydrogen production, certain plants are deprived of oxygen and sulphur. When this occurs, the plant produces hydrogen. Various algae, for instance, contain the enzyme hydrogenase, which splits water into hydrogen and oxygen. There are literally hundreds of primitive plant forms, such as lichens, which can produce hydrogen.

There are numerous other experimental techniques that are being evaluated for hydrogen production as well.

Types of Fuel Cells

Above, PEM Fuel Cell,.

Below, Alkaline Fuel Cell

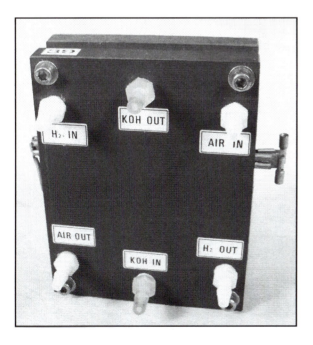

There are five basic types of fuel cells:

AFC   alkaline

PEMFC   proton exchange membrane or polymer electrolyte membrane (either name is correct)

PAFC   phosphoric acid

MCFC   molten carbonate

SOFC   solid oxide

Fuel cells are named or defined by their electrolyte, i.e.: phosphoric acid, molten carbonate, solid oxide, or proton exchange membrane. PEMFCs have a solid ion exchange membrane made of a sulfonated Teflon-like material (which is the electrolyte), and platinum catalysts. SOFCs use solid yttria stabilized zirconia as an electrolyte, and perovskites as a catalyst. PAFCs use a liquid phosphoric acid as an electrolyte, and platinum catalysts. MCFCs use a liquid alkali carbonate mixture and nickel catalysts. AFCs use a liquid potassium hydroxide catalyst and platinum electrodes.

Operating temperatures

All of these fuel cells have different operating temperatures. For instance, alkaline cells run from 50-200° C, PEM cells from 50-100°C, phosphoric acid run at about 220°C, molten carbonate run at about 650°C, and solid oxide run at about 500-1000°C.

Generally, higher temperature cells are preferred for stationary applications and lower temperature cells for mobile applications. PEM cells will probably be the choice for both home power applications and vehicles. They have a low operating temperature and are convenient because of their solid electrolyte. They are easy to handle, service, and maintain.

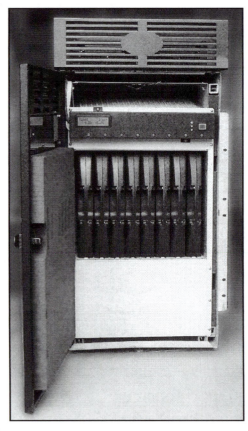

*Commercial PEM
fuel cell unit*

Regenerative fuel cells

In passing, I should also mention "regenerative" fuel cells. Such a cell acts as both an electrolyzer that produces hydrogen and oxygen gases, and as a fuel cell, using these gases to generate electricity. This is not a new idea – in fact, Grove's first fuel cell was a type of regenerative fuel cell.

System and Components Overview

Photovoltaic Fuel Cell Systems

A photovoltaic fuel cell system consists of a low voltage, high current solar panel (also known as an Electrolyzer Specific Photovoltaic Module – ESPM), an electrolyzer, gas scrubber, hydrogen and oxygen storage vessels, and fuel cells.

During peak sunlight hours, ESPMs produce electricity to power the electrolyzer and produce hydrogen and oxygen which can then be stored. Both hydrogen and oxygen storage can be in low, medium or high pressure tanks. Hydrogen can be stored in hydrides, or metal organic frameworks, or carbon nanotubes. Whenever needed, the hydrogen and oxygen is released to the fuel cell and electricity is generated.

Fuel cells vs. batteries

There are advantages in the use of fuel cells rather than lead acid batteries for electrical storage. Fuel cells are much lighter than batteries, and do not contain any heavy duty toxic substances like lead. And, the main advantage is that hydrogen can be stored indefinitely, so during the summer when sunlight is abundant, hydrogen can be made for use during the short days of winter when there is less sunlight available.

Components of a PV fuel cell system

In a classic PV power system there are two major components: the PV panels (Battery Specific Photovoltaic Modules – BSPMs) and the battery storage device. In a PVFCS (Photovoltaic Fuel Cell System) you have the PV panels (either BSPMs or ESPMs), the electrolyzer, storage tanks and fuel cells. General maintenance for such a system consists of making sure the water is refilled for the electrolyzer and checking on the plumbing now and then, and keeping your PV panels clear of snow, if you live where it snows.

Wind or small hydro power

A fuel cell system can also run from an electrolyzer powered by a hydro or wind source. A hybrid system with wind and photovoltaics also works well; and one could also use a BSPM system that recharges batteries and then shunts over to a bank of electrolyzers when the batteries are fully charged.

Add-ins for the system

In addition to the ESPM (or BSPM), electrolyzer, storage tanks, and fuel cells, one may also need an inverter to change the output of the fuel cells from DC to AC to run appliances; or one can opt to run DC appliances and skip the inverter.

A voltage regulator or DC to DC converter might also be desirable. The voltage regulator simply regulates the voltage out of the cells and the DC to DC converter delivers a certain voltage and current within defined parameters. DC to DC converters can be useful to step up low voltage from the cells to a more useful voltage for your applications.

Fuel cell configurations

Fuel cells, whether in stacks or individual cells, can be hooked up in series or parallel to either make more voltage or more current depending on one's needs. Most individual fuel cell slices produce about .5 volts at around .5 to 2 amps at full load. Larger cells will produce the same voltage with greater amperage, depending on the size of the active area of the cell.

There are practical physical limits to stacking even though theoretically one can stack fuel cells to produce whatever voltage is desired. For instance, a 25 slice fuel cell that produces 2 amps of current and is connected in series will produce about 12.5 volts at 2 amps current. Two such stacks connected in parallel will produce 12.5 volts at 4 amps; or 2 stacks of 25 cells each with each stack connected in series will produce 25 volts at 2 amps, etc.

Matching the PV panels to your needs

When contemplating a PVFCS, consider the type of PV panel you are going to use. You can purchase commercial panels or build your own. If you build your own, you can customize the panel output to match your electrolyzer needs. And, putting together your own photovoltaic power supply for your electrolyzer is a very cost effective way to go.

As a bare minimum, I use panels that are designed to give out 3 to 6 volts with 10 to 20 amps apiece because the electrolyzer needs very little voltage, but does need more current than the average PV panel puts out. (PV panels can be connected in parallel to generate more current to power more electrolyzers.)

Building your own panels also allows you to take into consideration your local climate. In New Mexico, for instance, there is much more sun than in Vermont where there are many cloudy days. If I am building panels for the Vermont climate I can add more solar cells to the panel configuration so that with less sun I can generate the minimum voltage and current needed to keep my electrolyzers working at the pace I need. These ESPMs output more voltage and amperage to take best advantage of the relatively low amount of sun that we do get.

Powering an electrolyzer

Although electrolysis starts moving at around 1.49 volts, I use 2.5 volts as the low threshold. Consider also the type of electrolyzer you are using and its minimum power input needs. For fuel cell power applications, figure out your power

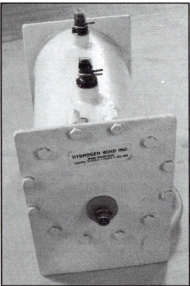

Commercial general purpose electrolyzer

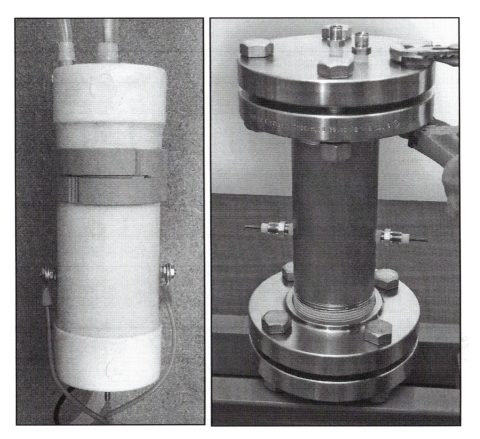

Two of the author's electrolyzers, designed specifically for intermittent power sources such as photovoltaics and wind.

requirements and how many electrolyzers you will need and then build your panels accordingly.

If you want to build an electrolyzer for high power fuel cell applications, I recommend my **Build a Solar Hydrogen Fuel Cell System**, which has complete details for building ESPMs and a planar fuel cell stack, as well as instructions for configuring BSPMs to power electrolyzers; and **Practical Hydrogen Systems: an Experimenter's Guide** which features a more sophisticated stainless steel electrolyzer. If you are considering building a PVFCS both of these titles will be very helpful for setting up the components for a PVFCS in a safe and efficient way.

Building an electrolyzer

You can build your own bench top electrolyzer. When building an electrolyzer, you can use nickel iron electrodes salvaged from an Edison battery. (Edison batteries can be found at industrial auctions or from surplus suppliers.) Nickel sheet can also be used as I did in the lab electrolyzer described beginning on page 72 of this book.

Compressing the hydrogen

If you want to pressurize your system and go beyond the lowest pressure storage, include enough power for a small hydrogen compressor when you calculate your photovoltaic panel needs. Many experimenters use propane fittings and tanks with some modifications to store their hydrogen. Pressurized storage and delivery systems are a topic that we will not cover here, but you can get additional information in my book **Practical Hydrogen Systems: an Experimenter's Guide** on this subject. For an experimental start I would suggest you begin with a low pressure system such as described in **Build a Solar Hydrogen Fuel Cell System**.

If you are serious about building a PVFCS system, start small, take your time, take notes, observe, and slowly build up your system.

More resources

For good information about small solar electric systems I suggest **Solar Electricity: A Practical Guide To Designing And Installing Small Photovoltaic Systems** by Simon Roberts (currently out of print, but worth the effort to get a copy). Also peruse the websites of solar suppliers like Backwoods Solar Electric Systems. These resources are full of valuable information about setting up and running PV power systems; and about components and circuits that are needed to do a variety of tasks related to PV systems. They will also give you a good idea of how to calculate and size your PV system for your needs and climate.

Building a PV panel for your electrolyzer

If you are interested in building your own custom solar panels to power your PVFCS, my book **Build A Solar Hydrogen Fuel Cell System** has complete details for ESPM construction. In brief, you can make the panels electrolyzer specific by connecting the PV cells in a parallels-series format. Instead of connecting all the cells in series, connect the cells to each other in rows in parallel, and then connect these rows to each other in series to make a lower voltage, higher amperage panel. As an example: for cells that output .5 volts at 2 amps, I can connect 5 cells in parallel to output .5 volts at 10 amps.

When cells are connected in parallel they add the current (amperage) from each cell. If you construct another row of parallel connected cells and then connect that row in series with the first row you will now have added .5 volts and your output from these two rows will be 1 volt at 10 amps. You can then add as many rows as you like to add more voltage. Ten rows in this configuration will give an output of 5 volts at 10 amps.

To increase your amperage per row, simply add more cells to the row. If you have ten 2 amp cells connected in parallel in each

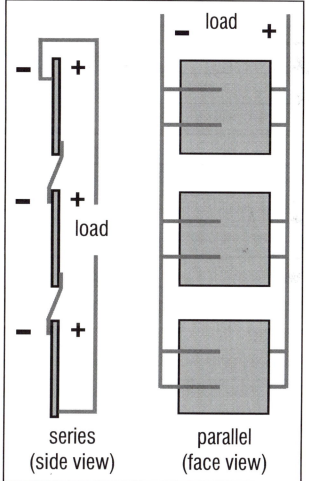

series
(side view)

parallel
(face view)

PV cell faces are negative. The backs are positive

row the output will be .5 volts and 20 amps per row. If you connect up ten of these rows in a series fashion you would have an output of 5 volts at 20 amps. Such a panel would consist of 100 2 amp cells. This can be constructed as one panel, or broken down into several smaller units connected to each other.

Figuring PV panel costs

Solar cells come in a variety of shapes and sizes and current (amperage) output. Cells that put out less current are less expensive, but then you have to use more of them to get up to the current you will need. Higher output cells are more expensive per cell, but you need fewer of them to get up to current.

When pricing your cells for this type of system, compare the cost of building the panels with lower output cells versus higher output cells. Sometimes it is cheaper to buy the higher output cells and sometimes it is cheaper to buy the lower output cells. The PV cell market is quite variable.

If you use lower output cells, you increase your costs for tabbing materials, as well as for framing and glazing materials for the larger panels. Be sure to figure in all the variables when calculating your panel cost.

It's also possible to buy used or discontinued solar panels. If the price is right, this option can be very cost effective. The minimum output to run an electrolyzer would be about 4 volts at 10 amps. Although it costs more, it is better to oversize as much as you can afford to, especially if you live in a low

Constructing an electrolyzer specific photovoltaic module

sun area. If you have an existing PV system, of course, you can simply connect to your system.

Wiring the system

In general, just as with a battery storage PV system, be sure that all wiring and/or bus connectors are rated to carry the current produced anywhere in your system, whether it is the connections from fuel cell stack to fuel cell stack or connections from the electrolyzer to the panel, etc.

At the very minimum, fuses should be added at various points in your system as a safety measure. Check the websites of solar suppliers such as Backwoods Solar Electric Systems for more info.

Support and use components

In addition to the main components of the system – PV panel, electrolyzer-scrubber, storage tank, and fuel cells – there are support and use components to consider. For instance, when the storage tank is full, you need a safe switch that shunts the current from the panel away from the electrolyzer so that it stops producing gas. You will also need a visual pressure indicator, vent valve for the storage tank, and flashback arresters. These components are basic to any type of safe system. Also, it's a good idea to ground the components for static discharge, induced currents from nearby electrical activity and direct discharge of lightening.

Housing the system

The housing for your system should have proper ventilation so no accumulation of gases can occur. A one-side-open shed with a vent in the peak of the roof will protect the system from the elements, and answer to the venting problem. Hydrogen rises and dissipates quickly if it has somewhere to go.

Control valve

You will need a control valve to let the gases flow into the fuel cells when you want power, and to cut the gas flow to the cells when you want power generation to stop. The control valve can be automatic, remote or manual, or a combination of these. You may also want a switch to connect and disconnect the fuel cell output into the wiring, distribution panel, inverter, or DC to DC converter.

Supercapacitors

Supercapacitors can be used in parallel connection with electrolyzers and/or fuel cells to aid in a smoother delivery of either gas output or power generated from the fuel cells.

Supercapacitor connected in parallel with a fuel cell

Solar powered electrolyzers reduce output with quickly changing cloud transits. A supercapacitor allows smoother, more consistent gas production.

Coupled in parallel with fuel cells, supercapacitors provide a buffer for a fuel cell or fuel cell bank when current surges are needed to start electromagnetic devices such as motors. For more information about supercapacitors, see my **Solar Supercapacitor Applications.**

PEM Fuel Cell Components

Building a fuel cell is easy. Anyone can construct a commercial quality PEM fuel cell with some very basic tools and materials. There are very few components that make up the fuel cell, and the materials are easy to access.

The heart of the fuel cell is the MEA (membrane electrode assembly) which is sandwiched between graphite plates that act as electrodes and gas flow fields. On either side of the cell are the end plates. In detail, a PEM fuel cell consists of the MEA, graphite plates that serve as both electrodes and flow field plates, electrode conductors, rubber gaskets, end plates and barbed hose connectors to accept tubing for gas flow, a few screws to hold it all together and some binding posts for electrical connections.

The MEA is composed of a proton exchange membrane with a carbon cloth or paper diffuser loaded with platinum catalyst on either side. There are several ways to make MEAs.

In planning the process of hands-on research for this book, I decided to begin with purchased MEAs in order to gain familiarity and confidence with the process of fuel cell design and construction, rather than begin at a more "from scratch" level which requires more equipment, and more initial expense for materials.

However, if you want to produce a lot of cells at a reasonable cost, then be sure to read the chapter about making experimental MEAs, and look at the resource section of this book to find more details. In the process of learning to design and construct fuel cells, I discovered that to produce an MEA is actually quite simple, and not at all as intimidating as it first seemed.

MEA basics

The first component needed for an MEA is a polymer electrolyte membrane such as Nafion. Then you need a carbon or graphite fabric or paper that will serve as the gas diffuser. Finally, there is the platinum catalyst which is deposited on the carbon fabric or paper. Getting the platinum onto or into the gas diffuser can be accomplished in a variety of ways, including sputter diffusion, electroplating, or using inks or powders with carbon black or porous carbon and platinum in them.

Once coated, a piece of the platinized fabric or paper is adhered to either side of the polymer electrolyte membrane by adding a liquid polymer electrolyte solution and hot pressing the layers together.

Purchasing MEAs and components

These materials can be purchased in various states of finished product. For instance, platinized fabric can be purchased, or you can apply the platinum catalyst to carbon cloth yourself.

In the next few years MEAs will probably be produced in sufficient quantities (much as solar cells are today) so that the prices will drop to the point where it will be cheaper to buy them than to bother to make them.

Surplus fuel cells

Also, within the next few years, surplus fuel cell units will probably become available at very reasonable prices as the military and other government institutions begin to discard their equipment for newer models, etc. Auto makers will soon begin to use fuel cells for vehicles and surplus units will become available through this industry as well.

As more fuel cells of all kinds become available as surplus, there will be more people experimenting with these units in home power and other applications. I believe that within a short period of time fuel cell units will become almost as common as batteries. And, despite the fact that this technology is over 170 years old, there is still plenty of room for the experimenter to come up with new designs and ideas. As I will show you, fuel cells are a simple technology that can be mastered with minimal tools and limited resources. (For the simplest practical fuel cell to make, see the chapter about my graphite foil fuel cells – no power tools are necessary for their construction!) I hope that this book will inspire you to get tinkering and invent something interesting!

Building the K18 Hard Graphite Convection Single Slice Fuel Cell

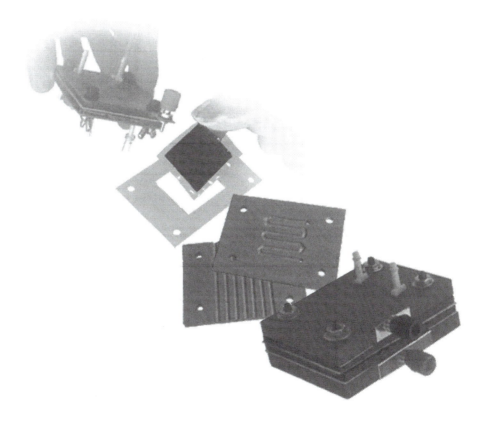

K18 Fuel Cell - Tool List

Band saw, wet & dry such as the DB-100 from Inland Products distributors.

Diamond blade for band saw. If you purchase an Inland Products DB-100, it comes with a diamond blade; however you should check on this before purchase.

Drill press. Any low cost, small drill press will do. You can purchase these for about $60.00. You might also be able to use a Dremel tool. Supplier: Northern Tool and Equipment or local hardware store.

Drill bits. Three drill bits, a $7/64$", a $9/64$", and a $7/32$". These can be purchased at your local hardware store

Router bit. $1/8$" router bit from Micro-Mark, part #60719.

Tin snips. From you local hardware store.

Pliers. Available from your local hardware store

C-clamps. Available from your local hardware store

Squares. Available from Micro-Mark, or local hardware store, etc. If you do not have a fence on your band saw or drill press, you will need a square to align whatever you use as a fence.

Metal break (optional). Available from Micro-Mark. Part #82817, or local hardware store.

Fences and angle plates. Angle plates guide your work on the band saw and drill press. You can make your own or purchase them. For instance, the Micro-Mark angle plate, part #60626 can be used as a guide when working on the band saw or drill press. Your local hardware store or Northern Tool has items that meet this need. Angle plates are optional, but a fence is necessary. Fences are available at a variety of sources or can be made.

File or grinder. For smoothing parts if necessary. Available at local hardware store and other sources.

Sandpaper. Use whatever you feel is necessary as far as grit. You can use 200 grit plasterer's screen if your surface has a lot of uneven marks, and then graduate to a 320 and then to a 400 grit. You may only need a 400 grit if you do your graphite slabbing well.

Exacto knife, or utility razor knife, or razor blades, and scissors.

Fine point felt tip marker. Sharpie or similar marking pen.

Carbon paper if you want to use templates.

Multimeter or voltmeter and ammeter to read voltage and amperage. Can be purchased at local electronics store such as Radio Shack. Be sure to get a multimeter, ammeter that can read up to 10 or 20 amps.

Hole punch. Small-diameter hole punch, $7/32$" hole diameter, trade size 8. Part #3424A18, McMaster-Carr.

K18 Fuel Cell Materials List

End plate material. Electrical grade fiberglass sheet (GP03), $3/8"$ thick, 12"x12", red, part #8549K47, McMaster-Carr. You can substitute other material for this, such as $1/2"$ thick PVC sheet, McMaster-Carr part #8747K116.

Graphite for gas flow conductor plates. Some of the suppliers offer blanks cut to size, or you can order pieces to cut yourself (be sure they fit in your band saw). Local graphite companies may have discarded pieces that can be obtained quite cheaply from their scrap piles. See resource list.

Silicone rubber for gasket and spacer material. Thin gauge silicone rubber .020" thick, 12"x12", 35A durometer, part #86435 K45. McMaster-Carr

Mylar sheet for membrane electrode assembly surround. Comes free as carrier sheets with the silicone rubber sheet (above).

Membrane electrode assemblies (MEA). See MEA suppliers, page 173.

Binding posts. Insulated binding posts, part #274-661, Radio Shack.

Sheet metal foil for electrodes. Nickel alloy foil, .010" thick, 4" width, plain, 1' length. Part #8912K24, McMaster-Carr.

Screw fasteners, nuts, washers. Four-10-24x$1^1/2"$ socket head cap screws. Four-10-24x$1^1/2"$ hex nut, course thread. Eight $3/16"$ flat steel washers.

Shrink tubing. $1/4"$ that shrinks to $1/8"$, or whatever you can find that is close to that. Local hardware stores, electrician shops automotive supplies and Radio Shack are good sources.

Barb splicer hose connectors. Can be found at well stocked hardware store. Small Parts Inc. has them available as well as other sources such as McMaster-Carr. You will need two $1/8"x1/8"$ barb splicer connectors. Get more than you will need so that you will have extra on hand.

Silicone rubber adhesive sealant. Available at local hardware stores

Contact information for the suppliers mentioned can be found on page 172.of this book. For resources local to you, check out electronics suppliers like Radio Shack, and hardware and plumbing supply stores.

About the Materials

Buying MEAs and components

As previously mentioned, you can choose your level of do-it-yourself comfort for fuel cell construction. If you are new to fuel cell technology then, I suggest purchasing a PEMbrane (see MEA Suppliers).

Shop around to see what is available and what the best prices are. Note that if you are going to build a fuel cell stack, some suppliers offer a substantial price break for MEAs in quantity.

You will be purchasing membrane electrode assemblies without the mylar surround. You can easily add the mylar surround yourself.

The listed suppliers have other membrane sizes available – contact them for prices and availability. Some also sell the various components for MEAs if you wish to get closer to starting from scratch; and some provide other components for building fuel cells if you do not wish to make them yourself, such as graphite conducting flow field plates. It's also possible to buy complete fuel cells and small electrolyzers for powering them.

Mounting the MEA

The MEAs you purchase will have a margin of very thin transparent PEM material on all four sides. To prepare your PEMbrane assembly for insertion in the fuel cell, this margin will be sandwiched between two layers of mylar frame, or "surround." When you purchase your silicone rubber

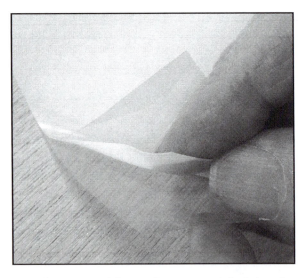

Silicone rubber sheet, sandwiched between mylar carrier sheets

(from McMaster-Carr), it comes sandwiched between what is called carrier sheets. These are made of mylar and can be used to make the mylar surround needed to complete the membrane electrode assembly.

Rubber gaskets and spacers

The rubber sheet is used for the gas gasket to seal the edges around the hydrogen flow field plate, and also for compressible spacers between the graphite plates and end plates. At first I was going to use latex but decided in favor of silicone as it has better characteristics for the operating temperatures of a fuel cell.

Keep the components clean

The silicone rubber sheets that you will purchase for gaskets and spacers pick up dirt and dust quite readily, so it is important to keep the material clean. I will add here that after preparing the various parts for your fuel cell, it wouldn't hurt to wash everything in distilled water for a final cleaning before assembly. However, make sure the pieces are dry before assembly.

Working with graphite

Graphite plates (at least for this design) provide a gas flow field and act as an electrode conductor. It is an interesting material to work with and is quite easy to machine, although a bit messy. It is the most stable form of carbon and is found naturally as well as being manufactured. Much of the graphite used today is manufactured.

Graphite has a melting point of 6,600° F, is basically chemically inert and electrically conductive with a resistivity of about 0.006 ohm -cm. This makes it a good choice for many applications including fuel cells. The name "graphite" comes from the Greek word for "to write" as it is a component of lead pencils. It is slightly flexible but not elastic and its color can be described as a black silver or grey. Because of its softness, It also makes a good lubricant.

Obtaining graphite

Graphite for this project can be purchased from any of the suppliers listed. If you purchase graphite from secondary suppliers, be sure to specify what type of pieces you prefer. Purchase square pieces with no wedge shapes unless you have the shop skills and equipment to square off odd shapes. The blocks should not be bigger

Graphite blocks ready to slab

than your band saw can accommodate. I will discuss that in more detail later. You will have to slab and drill the graphite as well as machine it with a router bit on a drill press.

End plates

The end plates are cut on a band saw from a piece of fiberglass sheet. Other materials such as 1/2" PVC sheet work well, too. The important thing about end plates for a fuel cell is that they must be rigid throughout their length and height, so that when the screws are tightened on each of the four corners of the fuel cell, the end plates will provide even pressure throughout their surface. This ensures that the graphite conducting plates have good contact with the MEA. End plates with too much bend will not translate the pressure exerted by the screws to the center of the piece where the MEA is located.

To put it in a nutshell, a rigid material is needed for end plates. I have not had a chance to experiment with many different materials but anything can be used as long as it is rigid, non-conducting, and will hold up to moisture. There are many appropriate plastics for this application. Make sure to check the material's specs for ease of cutting and drilling.

Metal plates can also be used, as long as they are insulated from the graphite plates. On the other hand, if the end plates are used as conductors they do not have to be insulated from the graphite slabs. I have seen fuel cells with metal plates, some plated with gold in order to provide the best electrical conductance.

Miscellaneous parts

Other materials needed are nickel sheet metal for electrodes and a few screws, nuts, washers, some shrink tubing and several small pieces of Mylar cut and punched to act as membrane holders.

Fuel cell jazz

It might be an interesting exercise to build a fuel cell from common found materials as much as possible just for fun. After you begin to work with some of the materials in this book and you put together a fuel cell, you will no doubt think of ways to improvise as you begin to understand how you can use other materials and come up with better, less costly and more efficient designs.

The materials used in the construction of the fuel cell in this book were chosen intentionally so that the end product would be a commercial quality fuel cell that would last for many years and provide high performance characteristics under the most rugged conditions, without need for repair due to material failure. Other important considerations were cost, and ease of construction so that anyone who has access to fairly common tools could easily construct a fuel cell.

Simple Tools Do the Job

The only power tools needed to construct a fuel cell from hard graphite are a band saw with a diamond blade, and a drill press. The band saw is needed to slab and cut the graphite plates and cut the fiberglass end plates. The drill is needed to drill holes and rout the flow field patterns into the graphite plates, and to drill holes in the fiberglass end plates.

Wet and dry band saw

The band saw should have a variable speed control and be a wet and dry saw. If you have a band saw that is not a wet saw, you can rig up a water dripper device so that water coats the substances being sawed. A wet saw simply has an attached water reservoir with some tubing that ends above the cutting area so that water can drip on the material as it is cut. The drip rate is usually controlled by a screw that pinches the tube to let more or less water out depending on

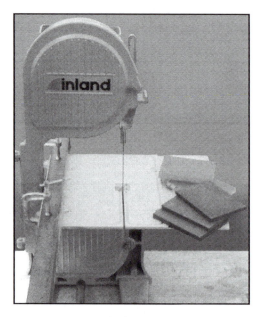

Wet and dry band saw

the thickness and the material of the work piece. The water acts as a lubricant and helps keep dust out of the air. Some materials need to be cut wet, some need to be cut dry, and some can be either.

Instead of the band saw, you could use a lapidary saw with a diamond cutting blade, or a regular table saw with a diamond blade. The blade would have to be wide enough to cut the height of the fuel cell. I did look into it briefly, but did not find lapidary saws that could accommodate the height. I also checked into tile saws but again they did not have the height needed.

Another possibility would be to use a hacksaw with a diamond blade and a very good miter box. Just the thought of it hurts my arms, but I don't see why it wouldn't work with a good miter box that had the appropriate height to hold and cut your piece. A local stone quarry or other business that cuts stone might be willing to cut the pieces for you.

Drill press

The drill press should have the option to move the belts in the head down to a lower speed for cutting graphite (about 500-700 rpm). That is probably the best speed range, but a little higher may not hurt – it's just not quite optimal.

Instead of using a drill press, you may be able to use a hand drill with a special jig you can buy from the hardware store to drill straight holes. The drill sits in a small frame and can then be used like a drill press. This would require a hand drill with a slow speed selection or variable speed control. A Dremel tool might also be used to both drill and rout. I haven't tried either of these so I don't know for sure how they would work for these purposes. The Dremel tool would also need a variable speed or slow speed setting.

Setting up for fuel cells

If you are going to make several fuel cells or more, you may want to set up some permanent jigs for your drill press and band saw so that you can speedily and easily form your parts. If you have access to a milling machine, of course that would be the best way to machine your fuel cell parts, but not everyone has the cash to outlay for such machines. One of the purposes of this book is to provide information so that the average person who is not a tool wizard could make these parts easily and inexpensively.

In regard to power tools – use whatever works! I have used a band saw and drill press for the cutting, drilling and routing the materials, and they worked well. You may find easier ways or better tools and materials for making fuel cells than suggested here. By all means go for it!

Preparing the Graphite

Graphite is extremely easy to machine but it is messy. A respirator or some type of mouth and nose covering should be used to avoid inhaling the graphite particles while cutting and drilling. When handling graphite you will notice that it has a slightly greasy feeling and that it rubs off on the hands and clothes quite readily. So, when you work with this substance, wear clothes that you don't mind getting graphite on. Also, and most important, cover the motor housing on the bottom of the band saw so that no water or graphite can get into the motor. Graphite is conductive and will short out the motor – so, cover the motor very well.

Buy graphite pieces that fit your saw

The band saw that I use has a height clearance of a little over 3", so when I order graphite I have to make sure that at least one side of each graphite piece is no more than 3" in order to fit into the band saw. Measure your band saw height clearance and make sure the graphite pieces you purchase will at least fit under it to begin your initial cuts, otherwise you will not be able to cut the graphite with your band saw.

Setting the water flow

When slabbing with the wet dry band saw, use the drip control on the plastic tube (a screw that compresses the tube) to control the water flow over the work piece. The thicker the piece you are working, the more water you will need for the cutting surface. Thin pieces require less water. You do not need an excessive amount of water dripping on your work piece. If you see powder forming, you need more water, but if you use too much water, you will get spray and puddles of water on your work surface. Some band saws come with a leather wiper to keep the blade clean and wipe away excess water.

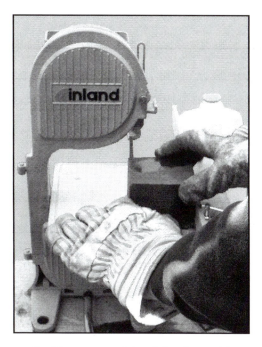
Cutting graphite blocks with wet band saw

Slabbing the graphite

As you slab with the band saw, do not push the work into the saw blade with any great pressure. Pushing the material aggressively into the blade will not increase the cutting speed. Just use light pressure to guide the material into the blade — let the blade do the cutting, don't force it. It will gently and slowly move along, removing material at its own pace. Read the instructions that came with your band saw, if you have them. Slabbing the graphite really is very easy once you have practiced on a piece or two. The slower the speed, the better the cut you will have.

Begin to cut

The first task will be to cut the block into 3/16" thick slabs with cuts as even and smooth as possible. When cutting, make allowance for the kerf of the blade.

Fences

You will need a fence to guide the material at the proper thickness for the cut. Your band saw may have one, but anything you can rig up to guide your material as you are cutting it at the proper thickness will do, or you can buy a fence. For this project I used a piece of angle iron c-clamped on to the work surface of the band saw and squared it to the band saw work surface using a square; however, that is not the most accurate way to go. Some saws have a groove in the table top that can be used as a guide for a home made fence, or you can use a 3 x 3" tool makers' angle plate (sold by Micro-Mark) to act as a fence if you can secure it with elongated c-clamps. Anyway, a straight piece of anything can be used if it can be squared with the band saw table, and if it can be clamped into position, and then moved for a variety of sizes of cuts. It should ideally be near the height of the piece you are cutting, if possible. You will save yourself a lot of trouble and aggravation later by taking the time to set up a good fence before you make any cuts.

Ripping gauges

You can use ripping gauges, though they are not necessary. They are essentially pieces of material the same thickness as the slabs you are going to make. Put them against the fence and when you adjust the fence, just push the fence with the ripping gauge up to the blade to get an exact measurement of the space you will need. Then clamp or tighten your fence, remove the gauge and you're ready to cut. Ripping gauges can be made or purchased.

Plan the cuts carefully

Take your time to think out the cutting procedure and to set up your fence. It is important to cut the graphite as accurately as possible. The more accurately you do this, the better the parts will fit together, and the better the fuel cell will work. Be sure to practice cutting a piece of graphite to get the feel of using the band saw, getting the drip rate right and feeling how to gently move the piece through its cut, etc. It's actually very easy – you'll become an expert in no time.

Cut extra pieces

When you make the plates, make more than needed for the fuel cell if possible. This way you'll have an extra piece or two in case something goes wrong with one of them later. Graphite, although easy to cut, is messy and once you clean the band saw after using it, you will not want to have to cut another piece and clean up the band saw a second time.

Err on the thicker side

Each graphite flow field-conductor plate is 3/16" thick. Avoid having it be any less than 3/16". If you can't get it right on, err on the side of your plate ending up being more than 3/16" thick. Remember to proceed slowly and evenly with a good fence and your cut should come out as smooth as is possible with this type of machine.

Cut the plate to size

Once the graphite is slabbed (cut into pieces 3/16" thick), you can then cut your plates to size, 3 3/8" x 2 3/8". After you have slabbed the pieces on the band saw, feel the cut surfaces with your fingers and you will notice that they are not smooth. You need a smooth surface, so the next step is to smooth out the cut surfaces of the plates with sandpaper.

Sanding

If there are deep ridges, start sanding with a plasterer's screen-type sandpaper. This is sandpaper or mesh that looks like screen – you can see through it. It will remove gross irregularities quite fast, so be very careful using it as it is very easy to take too much off.

Sanding surface and technique

To sand, you will need a very stiff surface that is absolutely flat, like a piece of rigid metal or other substance that you can lay the sandpaper or screen on. A flexible surface will not do because as you sand you would create dips or humps that you definitely don't want. Lay your sandpaper on the rigid sanding surface, then grab the graphite and move it across the screen or paper back and forth letting the sandpaper do its job without you pushing down or adding pressure. If this is too tedious or slow you can add pressure but be very, very careful because uneven pressure will create irregularities.

If the slabbing cuts were pretty rough, start out with a screen of 220 grit and be conservative about your sanding – check every few rubs to feel and look at the surface. Then move on to a 320 grit sandpaper followed by a 400 grit. You may want to test the finer grits first and skip the courser stuff. Be aware that you can create more problems by over-sanding than under-sanding. The silicone rubber seals will take up the slack of irregularities to some degree.

Perfection is not necessary here, just a fairly smooth surface without deep bows or humps. This process is easy to do and will only take a little time for each piece. Remember that if you make the best and smoothest cut on the band saw that you can, you will not have much sanding to do. It is a bit like building a house. If you start with a fairly good foundation (first cut) you will have to do less adjustment as the rest of the house is put together. However, don't worry too much if your graphite plates are less than perfect. It just takes a little practice working with this material to get a feel for its nuances. Anyway, I think you will be pleasantly surprised at how easy it is to slab and cut the graphite pieces you need.

Cleaning the band saw

Once the graphite plates are cut, the band saw will need to be cleaned. Not all the graphite slurry goes out of the drain tube and if you open the side door and look underneath, you will probably see graphite that has collected on the bottom and on the wheels. This should be hosed or washed out. Remove the motor from the band saw so that you do not get the motor wet, and then the rest of the machine can be hosed down or washed quickly and efficiently.

Using the templates

You can copy the templates in this book to mark the precise positions for drilling, routing and cutting the various parts needed for the fuel cell. The templates for the K18 can be found starting on page 179.

The next task in preparing the graphite plates is to cut the holes for the fasteners that will hold the fuel cell together.

For the two flow plates, cut the templates out (cut to the outside of the lines). Also cut 2 pieces of carbon paper to the template size. Put a piece of carbon paper on each graphite plate and align it perfectly. Place the template on top of the carbon paper and graphite plate (scotch tape is good for holding them in place). Use a pencil or a similar tool to transfer the template markings to the

Graphite plates, carbon paper and templates

graphite. Mark the crosshairs for the holes as well as the hole outlines so that you will be able to line up the drill with the center of the hole accurately. Then, mark the outlines and fill in the flow field patterns for both plates. The flow field pattern for the hydrogen side is a serpentine pattern. The flow field pattern for the oxygen-air side is a series of vertical grooves. When the template and carbon paper are removed, the exact positions for drilling the holes and routing the flow grooves will be marked on the graphite.

Drilling the holes

You are now ready to drill the holes in the graphite. Working with the drill press is fairly simple and straightforward. There are pulley settings on variable speed drill presses. Open up the top of the drill press and put the belt on the slowest speed pulley. Insert the 7/32" drill bit to drill the four fastener holes. To drill, you can rest your work against a fence clamped to the drill press, or gently clamp the graphite plate to the press table. You can also simply hold it in place with your hand while you drill, but it's a good idea to have the piece resting against a fence so that it does not move. When drilling the holes, go slowly and take your time – do not force the drill too much. Just let it cut easily. The holes are very fast and easy to cut.

Wear safety glasses when working with the drill press and watch your hands. Be sure that the piece you are drilling is secure and will not move or fly away. There should be safety and operating instructions that go with the drill press. If you read and follow them, you will be fine.

Routing the vertical flow fields

The next step is to rout the flow fields into the graphite plates. Start with the vertical oxygen side plate. Insert your 1/8" carbide router drill into your chuck. Set your table at the right height and set up a fence to accurately guide the piece in a straight line as the router drill cuts. Set the feed depth adjustment for 3/32". There are two nuts on the feed depth adjustment – be sure to spin down the second nut to lock the other one in place or the vibration from the machine may change the depth while you are cutting.

Routing the vertical flow fields

With the depth adjustment and fence set, position the plate so that the router bit is at the edge of the graphite plate, in line with one of the marked groove patterns. Then, turn the power on. As the drill spins, bring the drill down with the feed lever and gently push the graphite into the router bit, starting on one edge and going all the way to the other.

I am right handed, so I usually cross my arms – I bring down and hold the feed lever with my left hand while pushing the graphite plate through along the fence with my right hand. You can position the fence either to the left or right depending on what's comfortable for you. If you have someone to lend a hand, you can have them hold down the feed lever while you push through with both hands.

After you finish your first groove, reposition the fence so that the router bit falls into line with the markings for the next groove to be cut. I usually eyeball it and run the graphite piece up and down along the fence without the router running to see if it is positioned straight for the cut all the way through. Then I clamp the fence down to secure it. A square is handy to position your fence.

Each flow field groove is $^1/_8$" wide and the space between the grooves is $^1/_8$". When you are done you will have seven vertical grooves.

Routing the serpentine flow field

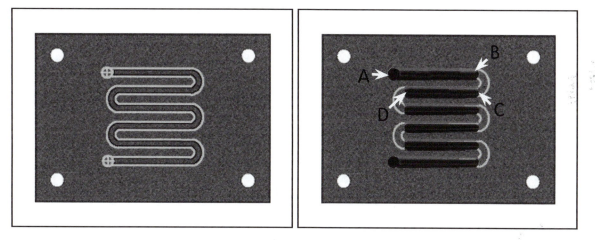

Copy the pattern onto the hydrogen flow field plate.

Rout the straight part of the flow field first, using a fence

The next step is to make the hydrogen flow field plate. Copy the pattern onto the graphite plate and then position the fence so that your router drill will begin its cut at point A. When you're set to rout, slowly bring the router bit into the piece as if you were drilling a hole. You will have set it at $^3/_{32}$", so when you reach that depth, begin to move the plate along the fence just as you did for the vertical flow field grooves. Stop at point B and then reset the fence to cut the next groove. Start at point C by drilling down to $^3/_{32}$", then move along the fence to point D. Begin and end the cuts just before the bend. Finish up the rest of the horizontals like this. When you are done you

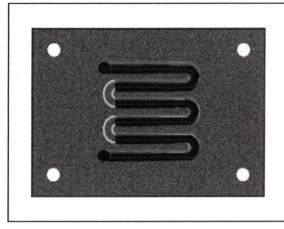

The curves to connect the straight grooves are routed freehand.

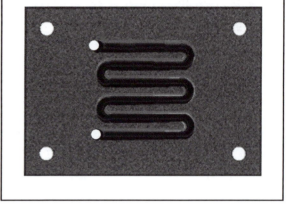

The gas inlet and outlet holes are drilled when the curves are complete

will have six horizontal grooves that are $1/8$" wide and spaced about $1/8$" from each other.

The next cuts, the curves to connect the parallel grooves and complete the serpentine flow fields, are made freehand. This would be tricky to do by yourself on a drill press. It's best to use both hands to guide the graphite as you rout the curves, but the feed lever needs to be held down while you are doing this. I don't advise using only one hand to guide the graphite for this task, so ask someone to hold the lever down for you while you cut the curves.

Go slowly and take your time on the curves. The graphite powder that you generate will hide the pattern marks, so you have to keep blowing it away as you cut so that you can see where you are going. I did all my hole drilling and grooving dry. If you can rig up something to wash water over the graphite and clear away the powder to see the groove curve better, then by all means do so. However I found that doing the routing dry worked well if I blew the graphite out of the way as I routed.

Cutting some curves on some scrap or test pieces before cutting the final piece is a good idea, to get a feel for it. These cuts are not hard to do – it's just a matter of getting used to working with the machinery and material. The inside of the curves can come out a little rough and may be less than perfect. You can smooth rough spots out a little with a diamond coated micro burr by hand. Just take one of the micro burrs and rub it on the inside of the groove. There are of course other ways to smooth these out if you need to. Do whatever works for you. Don't worry about it too much – the important thing is to have a channel for the gas to flow through.

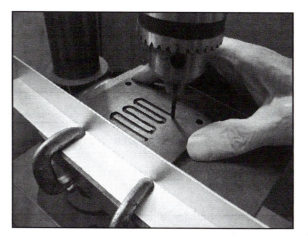

Drilling the gas inlet and outlet holes

If you really don't like doing these curves then you could probably connect the straight grooves with perpendicular grooves and 90° angled corners rather than curves. I have not tried this but it would be simpler in that it would give you a nicer cut and would be easier to do well. It might affect gas flow, though probably not enough to affect the performance of the cell.

When you have finished putting in your curves you can then drill the gas inlet and outlet holes with a ⁹/₆₄" drill bit. Use the template, mark off the inlet and outlet holes, position your piece, and drill.

Congratulations – you have just finished one of the most critical components of your fuel cell!

Completed graphite plates

Preparing the Other Parts

End plates

For end plates, I decided to use some
3/8" thick electrical grade fiberglass sheet
(GPO3) that I had on hand. It comes in 12"
x 12" sheets from McMaster-Carr. There
are many different plastics that you can
use instead of fiberglass – just be sure
that whatever you use can be drilled or cut
easily. The industrial supply catalogs listed
in the back of this book have many appropriate materials for this purpose. I chose
the fiberglass because I had it on hand,
it lasts forever and has excellent thermal
and electrical characteristics. However, it

*Cutting out the
fiberglass endplates*

*Drilling the fiber-
glass endplates*

is not the nicest material to work with. You definitely should wear gloves and a respirator when cutting and drilling fiberglass.

Insert a $7/32"$ drill into your press and drill four fastener holes in each end plate per the templates in this book. The holes will exactly line up with the four fastener holes in each graphite flow field plate. Next, drill the $9/64"$ gas port holes; then turn the pieces on edge and drill the $7/64"$ holes for the binding post screw into the top sides of the plates according to the templates. This completes the end plates.

Gas gaskets and spacers

If you ordered your silicone rubber from McMaster-Carr, you will notice that it comes sandwiched between what is called a carrier, which is simply two pieces of Mylar on each side of the rubber. Remove the carriers from the rubber and put them aside as this Mylar will be used later for making the frame for the PEMbrane. Keep the rubber clean as it tends to pick up all kinds of dust which sticks to it because of its surface nature.

The rubber gasket is thin gauge silicone rubber that is .020" thick and has a durometer rating of 35A. It has just the right amount of elasticity to compensate for minor surface flaws and unevenness in the graphite plates and provide a seal that will prevent gas leakage. This thickness should work well with the thickness of most purchased MEAs. However, for some membranes, including the ones you make yourself, you may have to use a different thickness so that your MEA carbon cloth contacts the graphite plate properly.

After you have removed the silicone rubber from the carrier sheets, place the templates for the rubber pieces on the silicone rubber sheet. There are several ways to cut out the rubber. You can place the template on the rubber and cut it; or you can mark it with a pen and cut it. You can use an Exacto knife, pair of scissors, rotary cutter, whatever. Take your time and cut the parts well – practice on a few pieces and see what works for you. Then, use the templates to punch the appropriate fastener holes and gas hole with a $7/32"$ punch.

Mylar surrounds for the MEA

Now you can take the Mylar carrier sheets that came with the silicone rubber and cut it according to the templates. You will need to cut and punch two Mylar surrounds. The center of each Mylar surround should have a square hole cut out, according to the template. You can cut the Mylar with an Exacto knife, scissors, or rotary cutter. If you wish you can also use a office-type flat bed paper cutter to cut the outside dimensions of the Mylar pieces. (This also works for cutting the outside dimensions of the rubber pieces.) Then, use the templates to punch the appropriate fastener holes with a $7/32$" punch.

Handling the MEA

Handle the MEA gently. It should always be handled with care. The transparent PEMbrane is very thin and delicate. It can be punctured easily, and a puncture can ruin it. Likewise, the oils from your hands and fingers are not good for the carbon cloth, so avoid holding the MEA by the carbon cloth, or use white cotton or nylon gloves to handle the MEA.

Mounting the MEA in the Mylar surrounds

Remove the MEA from the plastic bag. The MEA has a clear PEMbrane margin that extends beyond the carbon cloth on all sides. This will be the contact and gluing surface for the Mylar pieces that you cut out earlier. The MEA will be sandwiched between the two pieces of Mylar.

Assemble the MEA in the Mylar surround

Match the MEA with the square opening in the Mylar surround, to get an idea of how it fits. If the Mylar surround has been cut correctly, the carbon cloth should fit into the center hole with just a little gap all the way around between the carbon cloth and the edge of the Mylar. The clear PEMbrane margin will overlap the surface of the Mylar surround.

With silicone glue, draw a thin, continuous bead of glue around the square opening on one of the Mylar surrounds, but not too close to the edge. No glue should contact the carbon cloth when you glue the pieces together. After the glue is applied to the Mylar, pick up the MEA by the margins, center it and press the PEMbrane margins onto the Mylar. The PEMbrane margin should now be glued to the Mylar surround and the carbon cloth centered in the square opening in the Mylar. Press and smooth the glued pieces very carefully so that they are flat and even against each other.

The MEA mounted in the mylar surround

Next, with the unglued side of the MEA facing up, draw a continuous bead of glue along the PEMbrane margins, again taking care to not get glue on the carbon cloth. The idea is to make sure you have a good continuous seal. Place the second Mylar surround over the PEMbrane margins and press it onto glued area. Rub the seal with your fingers to press down and make sure the seal is firm. The MEA is now centered and glued between two pieces of Mylar surround, and ready for insertion into the fuel cell after the glue drys. Always let all the parts dry for at least 24 hours before assembling them into a fuel cell as a completed project. The edges of the Mylar surround can also be heat sealed by rubbing a low wattage soldering iron along the edges. If you do this, practice on some Mylar before you try it on your MEA. The edges must be smooth and have no burrs when finished. The MEA is now ready to be put in a fuel cell.

Electrode plate materials

Several different metals are suitable for making the electrodes. I chose a nickel alloy foil that was .010 thick as it would be easy to cut and drill and could also be used for electrodes in my electrolyzer. This way I could just order one piece of material for both purposes which cuts down costs. Stainless steel would be a good choice because it does not corrode easily. Any type of metal will work but some metals corrode much more easily than stainless steel or nickel. Copper, for instance, is an excellent conductor, but after a while the corrosion will impede the electrical contact – not good! The best contacts are made of gold as gold does not very readily corrode and is an excellent conductor. Many high end electronic components are plated with gold for this reason. If you have experience plating you may want to try this, however it is not at all necessary unless you plan for your cell to be made of only the very best, high end materials.

Cutting the electrodes

Cut the sheet nickel to size according to the template and drill the holes for the binding post screws. After you have cut and drilled the holes, put your electrodes in a small metal brake and bend the metal at the point indicated in the template. This will give you a nice clean 90° angle. Instead of the metal brake, you can sandwich the electrode piece between two stiff pieces of metal (or other very stiff material), line up the bend line on the edge of the sandwich, clamp the sandwich with c-clamps, then bend the protruding tab 90° and hammer it with a mallet to square it off. Whatever way that you can bend it and get a clean angle will work, and if you are working with the sheet nickel foil specified here, you'll find that it's quite easy to bend.

Cut the end off the barbed splicer

Gas entry ports

For the gas ports, nylon barbed hose connectors are inserted into the end plate. The type I purchased from the local hardware store is described as a $1/8"$x$1/8"$ barb splicer. This means that both ends of the connector have an outside diameter of $1/8"$. Notice that each end has a flared rim. This seals the connection when you place the vinyl tube over it.

Place your barb connector on a cutting surface and, with a utility razor knife, simply apply pressure and cut the flared rim off one side, which will leave you with a flared rim on one side and a smooth tube on the other. Put a tiny bit of epoxy around the outside rim of the tube and push the tubes into the $9/64"$ gas port holes on the hydrogen side end plate. You may have to twist and push as it should be a tight fit. Make sure that your epoxy does not drip or cover the holes as this would harden and definitely impede gas flow.

I was able to push and twist the connector in with my fingers, but you may find it easier to grip one end with pliers to get a tight grip as you push and twist. Depending on the barbed hose connectors, the fit may be just right or too tight. I had a few that bent as I tried to twist them in with pliers and I had to make the hole a little larger. In the photo you can see that the particular barb connectors that I bought had two tabs jutting out from the center of the connector. I used them as a stop block and pushed the connectors into the fiberglass end plate until the tabs were flush against the end plate. This left some of the connector protruding out the other side of the plate. With a razor blade, I simply cut off the protruding portion so that it was now flush with the plate surface.

Whatever type of barbed hose connector you use, trim it flush if a portion sticks out on the other side. Even if it is only a little bit that needs to be shaved, run a razor blade over it so that it is as absolutely flush as possible. If you don't find exactly the same type of barb splicer as I used, you should be able to find something that will work, as long as you have one flared end to accept the hose, and the smooth tube part is $1/8$" outside diameter to fit the hole in the fiberglass end plate.

Fasteners

To prepare the fasteners, you will need heat shrinkable tubing with an inside diameter of $1/4$" that shrinks to $1/8$" diameter. This item can be purchased at electricians supply, electronics supply, or industrial supply houses; hardware stores, or at a store like Radio Shack. The shrink ratio for the type of heat shrinkable tubing should be two to one. It's $1/4$" in its expanded state (the way it comes in the package). When heat is applied to it and it shrinks, it is referred to as being in its recovered state. Cut four pieces of the tubing, $1^{1}/8$" to $1^{3}/16$" long each. The shrink tubing acts as an insulator so that the cap screws are isolated from the electric current in the graphite plates.

Take four 10-24 x $1^{1}/2$" socket head cap screws. Put a $3/16$" flat steel washer on each cap screw, and then slip the heat shrink tube over the cap screws till one end of the shrink tube butts up against the washer. This will leave an end portion of the threaded section exposed. With a pair of insulated pliers, hold the screw with the shrink tube on it over a moderate flame. Do not let it get too close to the flame, otherwise you will burn the shrink tube. The idea is for the tubing to get just enough heat so that it shrinks to the size of the screw. Let it stay on the heat until you begin to see the outline of the threads through the shrink tube, but don't let it burn. The idea is to shrink it as tight as it will go – there will be a point where it will not shrink any further.

apply heat to shrink

You have now prepared all the pieces you will need for the fuel cell and are ready for assembly.

Assembling the K18 Fuel Cell

This is the easy part. Push the cap screws through the corner holes in the oxygen side end plate and lay the plate down with the screws sticking upward and simply put each part in its place, one atop the other following the diagram. If you have purchased a MEA rather than making it yourself, you will notice that one side is marked "anode". The anode side should face the serpentine hydrogen gas flow plate.

I designed this cell for tight assembly, so you may have to push a bit to get the screws through the end plates and graphite plates.

If you are having trouble fitting the shrink wrapped screws in, tap them in with a rubber mallet, or ream out the holes a very tiny bit so that you can more easily push in the screws. There should not be too much play in the screw holes as it is important for all the components to line up well. That is why a drill press is preferred over a hand drill without a jig – if the holes are drilled at an angle the screws will not line up well with other parts during assembly.

Don't worry if your parts are not perfect. If they are not too far off spec they will fit together with a little adjustment. The templates provided will help to keep everything on track and should make construction of the fuel cell parts much easier. It is important at each stage of construction to try to cut and drill each part as exactly as possible. This will prevent problems when it comes to final assembly. Take your time doing each piece – don't rush – and you should be pleased with your end product.

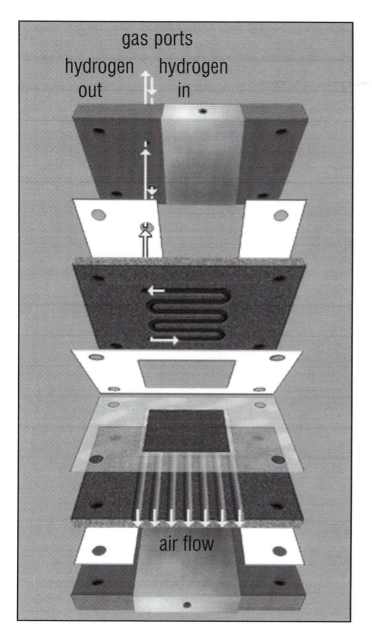

Layer order for the K18 fuel cell components

The completed K18 fuel cell

You now have a commercial grade single slice fuel cell that will last for an extremely long time.

Testing Fuel Cells

To test your fuel cells, you will need a multimeter or voltmeter, the fuel cell and a hydrogen supply. For a hydrogen supply you may choose bottled hydrogen, a metal hydride canister or an electrolyzer. Both the bottled gas and metal hydride supply can be a bit costly so an electrolyzer is probably the most cost efficient method to supply the hydrogen.

There are other methods for producing hydrogen in small quantities to test the cells, such as combining various metals and acids which react and produce hydrogen. I do not consider these metal-acid generators to be serious viable alternatives for on-demand hydrogen production for fuel cell testing. Preparing chemicals and cleaning up spent chemicals is too time consuming and messy to be a productive and pleasant experience while doing some simple testing

Building an electrolyzer

Building an electrolyzer is easy enough. The next chapter is a detailed account of how I built a bench top electrolyzer. Be warned that you must exercise appropriate caution when working with both caustic electrolyte and with electricity If you build this electrolyzer, you do so at your own risk. I discuss those risks in the electrolyzer chapter.

Buying an electrolyzer

The easiest method for obtaining a hydrogen source is to buy an electrolyzer (see the resource section of this book, page 169). With a purchased electrolyzers, you will have the least fuss and mess and they are easy to operate.

Hook up the hydrogen

That being said, we can move on to testing. Hook up the hydrogen supply to the gas inlet on the bottom of the fuel cell. You could connect it to the gas port on the top, but this cell is designed for gas entry on the bottom. Hydrogen is lighter than air and moves upward. If gas enters at the lowest port, if you have a minor leak, at least some of the gas will still pour over the membrane. This is a small point, since the seals should be adequate, but it doesn't hurt to over-design a bit.

Precautions

Before you turn on the gas supply to the fuel cell, be absolutely certain that there is no open flame in the area and that no one is smoking. Test your fuel cell outdoors. Do not have electrical contacts and switches nearby which can be turned on or off. They could produce tiny sparks that might ignite the hydrogen mixed with air.

Hydrogen is as safe as any other flammable gas, but as with any other flammable gas you must take proper precautions to ensure your safety. Actually hydrogen in some respects is safer than other flammables because it dissipates quite readily as it rises fast due to its lightness. It does not hang around and concentrate unless it's in a closed area where it can accumulate (for instance, where it's prevented from rising). Just remember that a hydrogen-air mixture is quite explosive and take all safety precautions that are required by any safety codes in your area.

Turn on the hydrogen

To test the fuel cell, connect the vinyl tube to your bottom fuel cell gas port. Turn on the fuel (hydrogen supply) If you are using an electrolyzer, it will take a few minutes for the gas to build up. Touch one of your multimeter or voltmeter probes to the hydrogen side graphite plate and touch the other multimeter or voltmeter probe to the oxygen side graphite plate.

Meter readings

The meter should be registering a voltage of anywhere from 1.25 v to .7 volts. At first you may get a higher voltage at about 1.2v, etc. and then you will notice that the voltage reading declines to .9 -.8 and may settle around .7 after a while. Typically these cells are rated at about .5

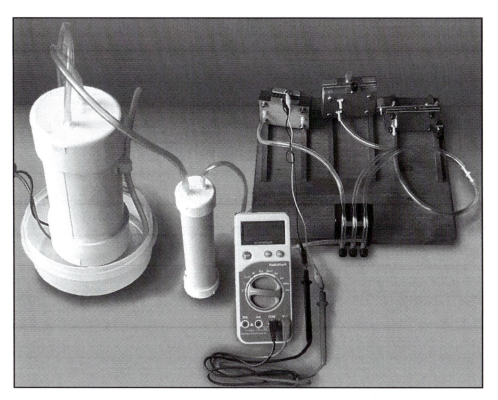

*From left, electrolyzer, bubbler, meter and
testing stand with three fuel cells*

volts at anywhere between .5 to 2 amps as they settle in with full load, depending on the MEA used. At any rate, if you are getting a voltage reading in this range your fuel cell is a success.

Test for continuity within the cell

After you have tested the graphite plate, put the probes on the metal contact electrodes and see if the voltage is being picked up there. The meter should have the same voltage reading. Then, put the meter probes on the binding post terminals and check the voltage there.

The purpose of this test is to make sure that the voltage is being carried from the plates to the terminals as it should be. If the same voltage does not register on the metal contacts as shown on the graphite plates, then you need to tighten the screws a little bit to improve the interface between the metal plates and the graphite plates. If the voltage does not show on the binding post terminal ends and it shows on the metal plates and graphite plates, then the connection between the terminal surface and the metal plate surface needs to be tightened.

Troubleshooting – no reading

If you do not get any reading at all then you have to systematically troubleshoot. First of all is hydrogen getting to your cell? If you are using an electrolyzer it will take it a while to build up enough gas and pressure, so patience is the key here. If that is not the problem, then the next likely culprit might be the graphite electrode flow field plates not touching the membrane surface adequately. The remedy here would be to tighten your screws a tad. Don't overdo it – proceed slowly. Oddly enough, sometimes backing up on the pressure will produce a better reading. This may be due to the fact that too much pressure from the fasteners constricts and reduces gas diffusion through the carbon cloth.

As you proceed with corrections, test the graphite plates with the probes. Fuel cells are really very simple and the most likely trouble spots will be that gas flow is restricted and not getting to the MEA; the seals are too leaky; or, the mechanical contacts are not touching tightly enough.

If all else fails in the testing and correcting procedure, then the MEA itself may be the problem. It could be that it is dud – highly unlikely, but possible. Be sure to test thoroughly – take your time before you reach the conclusion that you might have a dud MEA, as most likely it is something else. It is good to purchase at least two MEAs so that you have something to test against if you have problems.

Most likely, though, you will not have any problems with the cell. They are easy to construct and work quite readily even with a variety of design and construction flaws. The first cell that I made was thrown together rather haphazardly and despite that, it worked very well. More than likely you will not run into any problems, and you should feel free to experiment to see what produces the best reading.

Convection process

When you are testing and operating the fuel cell remember that it is a convection design. In other words, the oxygen side plate derives its oxygen from the air and operates on the principle of convection currents. The air reaches the MEA through the vertical slots. As the air is slightly heated it moves upwards through the slots, sucking in more air from the bottom. Thus a convection current is initiated and sustained. This technique works quite well and is less costly to implement than a cell which is designed to work on pure oxygen. However it should be mentioned that a pure oxygen intake might boost the fuel cells efficiency by as much as 30%. I will discuss this more in the chapter *Designing Fuel Cells*

So, when operating the fuel cell for testing or general use, it is best to prop up the cell so that the vertical air holes on the bottom are not resting on a flat surface. That would impede air from entering and thus impede the convection flow. Having air space under the cell is important for the correct functioning of the cell. If you are going to construct a bank of cells, you can make little racks or holders to keep things neat and convenient. Any rack or surface that you put the cells on should be made from material that is non conductive.

If you have made several cells they can be connected in series for more voltage so that you can run a small motor or light bulb. You could also connect them in parallel to produce more current. (See diagram next page.)

When you have finished building and testing your first fuel cell I think you'll be surprised at how easy it really was. You'll probably be ready to build more, make improvements and come up with your own designs for special purposes.

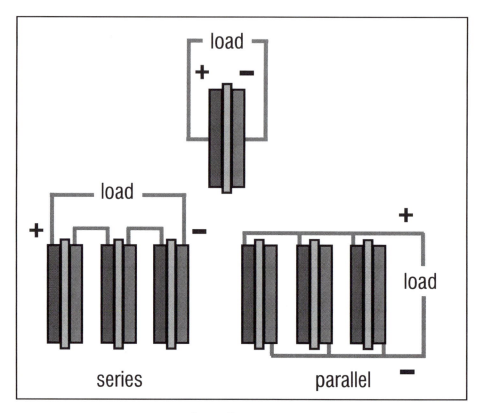

Single cell connections

Electrolyzers

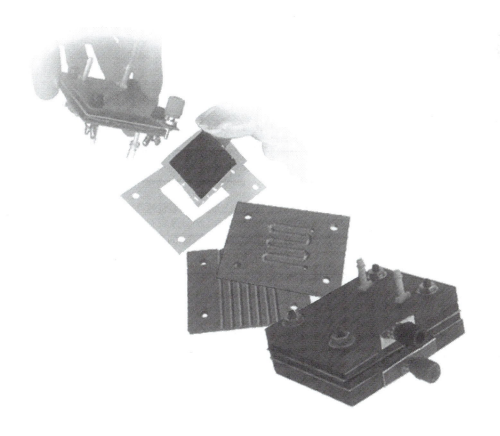

How Electrolyzers Work

An electrolyzer produces both hydrogen and oxygen gas for use in a fuel cell by the process of electrolysis. In the process of electrolysis, the atoms that comprise a water molecule, hydrogen and oxygen, are disassociated from each other. The disassociation occurs by application of electrical energy (a minimum voltage of at least 1.49 volts) to two electrodes immersed in an electrolyte solution composed of potassium hydroxide and distilled water. Distilled water is used as it has fewer impurities which would tend to collect on the electrodes and eventually inhibit and interfere with the electrolytic process. The electrolyte allows for ion conduction in the solution but is not used up in the process. Only the water is used and must be replenished.

Electrolytes

Potassium hydroxide (KOH), better known as lye, is a strong electrolyte. Electrolytes are classed as strong or weak. Strong electrolytes are, generally speaking, 100 percent ionized in solution. Weak electrolytes are much less ionized in solution. Strong electrolytes are better conductors of electricity and thus will enhance the electrolytic process by enabling more gas production per unit of energy input.

Electrolytes can be either an acid, base or a salt. Most soluble salts; sulfuric, nitric, and hydrochloric acids; and bases such as sodium, potassium, calcium, and barium hydroxides are considered strong electrolytes. Examples of weak electrolytes are acetic acid, nitrous acid, carbonic acid and ammonium hydroxide. Potassium hydroxide is the electrolyte of choice for electrolyzers.

Applying power

When the appropriate voltage is applied, hydrogen is liberated at the negative electrode, and oxygen is liberated at the positive electrode. The source for the electricity required for this process can come from a solar panel, a wind charger, hydro power, a battery, or a power supply. The current must be DC (direct current), not AC (alternating current). House current (AC) cannot be used in an electrolyzer unless it is rectified, as in a power supply that changes AC to DC.

In application, a minimum voltage of about 2.5 volts is a good starting point. Amperage (current) can vary depending on the volume of gas production you will need. To get a decent flow you will at least need one or two amps minimum.

Types of electrolyzers

The two most common types of electrolyzer are tank electrolyzers and filter press electrolyzers. Also, with polymer membrane technology it is possible to have an electrolyzer with a solid polymer membrane.

This polymer membrane is the same as used in the fuel cell, and, as a matter of fact, there are regenerative fuel cells that act as both an electrolyzer and a fuel cell. With a regenerative fuel cell, you can use it first as an electrolyzer and store your hydrogen; and then feed the hydrogen back to the fuel cell to generate electricity.

Grove's first fuel cell was a regenerative fuel cell since the electrodes acted as a source of current to disassociate the oxygen and hydrogen atoms in water molecules; and then when the current was turned off, the electrodes acted as a catalyst to recombine the oxygen and hydrogen into water again, producing electricity as a by-product.

The tank electrolyzer

The tank electrolyzer consists of a container with a positive and negative electrode that is filled with a 25% to 30% by weight solution of potassium hydroxide. The container (tank) has collection ports and fittings to conduct the gas to wherever it needs to be taken.

Home-built electrolyzers

There are many different designs for tank electrolyzers. As electrolyzers are easy and inexpensive to build, I decided to design and build my own for testing the fuel cells designed in this book. Also since I live in Vermont and heat with wood, I have an abundant supply of wood ash to produce my own potassium hydroxide (lye) which helps cut down costs for my experimental habits.

The electrolyte

Potassium hydroxide, also known as caustic potash, used to be a by-product of the timber industry. If you have an abundant supply of wood ash you should consider making your own electrolyte. If not, there are suppliers listed in the back of this book who offer potassium hydroxide, or you may find a local source.

Lye (potassium hydroxide) itself is very inexpensive, but since lye is a hazardous chemical you will have to pay a hazmat charge of about $20 plus the cost of the flaked or powdered lye itself and regular shipping costs if the lye is shipped to you. It's worth it to look for local sources of potassium hydroxide. If you live in the city there should be chemical supply houses that have it in stock, in which case, you can pick it up rather than paying the excessive hazmat delivery charge.

Most hardware stores carry sodium hydroxide (lye) rather than potassium hydroxide. I did find a product at a local hardware store that consisted of potassium hydroxide but it had another ingredient which makes it unsuitable for an electrolyzer. I suppose one could also use sodium hydroxide which is the most commercially available type of lye at the moment. Drano and Red Devil are the two most common brands of sodium hydroxide. I haven't used sodium hydroxide so I can't vouch for it as an electrolyte for these purposes. I also do not know what other ingredients these products contain, so again, I can't vouch for their use. I use potassium hydroxide as it is a better electrolyte and I have plenty of wood ash to make my own.

Preparing the electrolyte

If you purchase a commercial powdered or flaked potassium hydroxide, be sure to mix it with distilled water before you put it into the electrolyzer. When the lye contacts water it generates heat and may destroy your leak proof seals as the heat will stress the bonds. If you have ever poured lye down your sink to unplug a drain, you will have noticed quite a chemical reaction when it meets with the water. Wait until your lye solution cools down completely before you put it in the electrolyzer.

Safety precautions

Add the lye to your distilled water in small increments and wear total protective gear. Make sure your eyes, skin and body are covered and you can't get splashed. Lye, although a common household product, is a very caustic and dangerous substance. I have heard that 15 seconds of exposure to lye in the eyes will cause permanent blindness. It will cause burns on the skin also. If you work with lye, always wear safety glasses. I repeat, do not work with lye without safety glasses or a face shield. And you should always wear safety glasses at all times when you work around an electrolyzer, as well as protective clothing, that is, at least old clothing that will cover your body – long sleeved shirt, long pants, etc. Always wear protective rubber gloves on the hands to avoid contact with lye and always keep vinegar close at hand when you work with electrolytes as vinegar will neutralize lye. Keep a running water hose or a plentiful supply of water nearby.

Building the P38 Electrolyzer

Tools needed

Pliers.

Screwdriver.

Saw. Can either be power saw or any type of handsaw to cut PVC pipe.

Carbon paper.

Tin snips.

Hand drill.

Drill bits. $7/64$" and $1/4$".

Crimp Tool. Used for connecting ring terminals to wire to connect to the electrolyzer. Available at Radio Shack or any electricians' supply or electronics store.

Materials needed

PVC pipe 1½" inside diameter. Cut two pieces each 3½" long.

PVC pipe 4" inside diameter. One piece cut to 9½" long. Available at any hardware, building supply or plumbing supply house.

PVC end caps. Two PVC end caps for the 4" inside diameter pipe. (Outside of the pipe is 4½" diameter).

Binding posts. Insulated binding posts. Available at Radio Shack, part # 274-661.

Crimp-on ring terminals. Use ring terminals, not spade terminals. Ring terminals will tend to stay on whereas spade terminals can slip off, if the connection is not tight. Get a size that fits the 12 or 10 gauge stranded wire. Available at Radio Shack, hardware stores, etc.

10 or 12 gauge stranded wire. If you can get "zip cord" which your local auto store might carry, that would be good. Zip cord usually comes in black and red for DC circuits and is pulled apart to separate the wires – one black colored wire (for the negative input) and one red colored wire (for the positive input). The negative pole of the electrolyzer will produce hydrogen and the positive will produce oxygen. The two color wires will keep things clear and will prevent mixing gases if you disconnect and then reconnect the wires. Wire can be found at any hardware store, electrical supply, electronics store, etc.

Nickel alloy foil. This is the same nickel sheet used for the metal electrodes in the K18 fuel cell. If you constructed this cell, you should have enough left over for the electrolyzer electrodes. If not, it's available from McMaster-Carr, part #8912K24 nickel alloy foil, .010 thick, 4" width, plain, 1' length. You will be cutting two electrodes, each ¾" x 5¾".

Terminals for power supply end. You will need to connect the wire from the electrolyzer to whatever power supply you use with the proper terminal. Check the Radio Shack catalog and other electronic supply or hardware electrical section, etc.

Barb splicer, two ¼ x ¼". Available at Small Parts Inc., hardware stores and other suppliers.

Barb splicer, one ¼ x ¼" 90° elbow. Hardware stores and Small Parts.

Plastic tubing to connect to the hydrogen and oxygen gas port barb splicers, and the liquid level elbow barb splicer. Tubing can be purchased at any local hardware store.

Small plastic fastener to hold liquid level indicator tube in place on the side of the electrolyzer. Two-hole conduit straps can be used, and can be glued on, or secured with small screws. (Screws should not go all the way through the PVC pipe wall.)

Three small PVC couplers for feet. I glued them to the bottom cap. Anything can be used that will be stable and give enough clearance underneath the electrolyzer to accommodate the connectors, tubing and wire.

Screws and nuts. Two screws and two nuts for two of the holes on the electrodes. (The other two holes are for the binding posts.) The screws should fit snugly in the 7/64" holes that you have drilled in the pipe cap and electrodes.

Small plastic funnel (optional). A small plastic funnel can be connected to the end of the liquid level tube so that when the level is low you can fill the electrolyzer with distilled water through the funnel rather than taking the top off. The funnel can be epoxied to the tube.

Plastic (polyethylene) mesh screen. For barrier to keep electrodes from touching in the electrolyzer tank. You will also have to glue some plastic rails inside the tank so that you can slide the screen into the center of the tank have it held firm and in place. These plastic screens can be found at any craft store. They are used as substrates for needlepoint

You may find it helpful to make copies of the templates for the electrolyzer now, so that you can refer to them as you read the directions. There are 3 templates for the electrolyzer.

Purpose of the P38

Although the concept of the tank electrolyzer is basic and simple, the actual construction of a workable electrolyzer takes some planning and engineering expertise. I designed the P38 electrolyzer to be a bench-type electrolyzer to provide small quantities of hydrogen and oxygen for the purpose of testing fuel cells. It is not the most efficient or state of the art design, but my primary considerations were for lowest cost and sufficient quantities of gas for experimental purposes. I wanted something that could be made from readily available inexpensive parts, and with minimal tools.

The P38 was not designed to continuously produce gas for storage. Although the P38 will produce gas continuously – it is a real workhorse – it is not fitted to withstand pressure buildup for storage purposes. However, a simple modification in design would make the P38 suitable for a storage supply. If you are interested in storage then you should read my book Practical Hydrogen Systems: an Experimenter's Guide, which will give you more information about storage.

Use at your own risk

I designed the P38 for small scale production of hydrogen and oxygen for lab experimental purposes. It is not a design which is considered safe by any known safety standards. Therefore, be forewarned that if you build this electrolyzer and operate it, you do so at your own risk and knowing that it does not conform to any known safety standards: you have to take full responsibility for the use of this device. It can serve as a model for an electrolyzer that can be built within safety standards and codes once those standards are understood for your area and applied. You can also purchase an electrolyzer, either a PEM or potassium hydroxide type.

I have included templates that I used for the construction of the P38 lab electrolyzer. Although in theory electrolyzers are simple, building one that works well can be challenging. Whenever you mix electricity and water there is lot of potential for leaks, and thus, shorts. If you add a caustic electrolyte you are making the water more conductive and must make sure that all potential spots for leaking be eliminated.

Electrode separators

Although electrolyzers work with low voltage, the amperage is high, so never ignore any safety rules for working with electricity if you work with electrolyzers. Most tank electrolyzers have a separator between the electrodes. At one time asbestos diaphragms were used. The asbestos separator allowed the passage of the electrolyte back and forth but kept the gases separated and the electrodes insulated from each other.

With the P38 design I don't have to worry about the gases mixing, since I constructed it carefully, but there is no separator to insulate one electrode from the other. If the electrodes were to get bent up inside and touch each other there would be a short that would produce a lot of heat and probably pressure; and would certainly pop the top off and spew electrolyte everywhere – definitely not a good thing when we are talking about lye!!

A piece of plastic screen or polyethylene mesh is ideal as an electrode separator as it is fairly rigid and stiff. The screen can be cut to size and be inserted between the two electrodes, and held in place by gluing guide rails on the inside of the tank. It should not be tall enough to interfere with the gas collection tubes. If you design an electrolyzer you will want to add this feature. Again, I did not design this electrolyzer for other people to build and I am only including this information for educational purposes so that you can get an idea of how an electrolyzer works.

The P38 was quite simple to construct. I used a 9½" long 4½" outside diameter PVC pipe for the tank. The inside diameter of this tube is about 4". For the gas collection tubes that are epoxied to the inside of the top cap, I used two pieces of PVC that were 3½" long of what is called 1½" PVC pipe (1¾" outside diameter, 1½" inside diameter). The way the pipe sizes are marked can be confusing. At any rate, PVC is a readily available material – a criterion I try to stick with as much as possible when doing any projects.

A transparent electrolyzer

I also considered transparent PVC or acrylic or polycarbonate pipe which comes in a variety of sizes from McMaster-Carr and MSC. Transparent pipe can be valuable in that you can see all the action going on inside your electrolyzer. This is interesting and provides experimental observation opportunities, plus it's a good safety feature. Both of these suppliers have information on their web sites about the chemical tolerance, etc., of all of these materials so it's possible to assess their suitability for an electrolyzer.

Nickel electrodes

For the electrodes I used strips of nickel alloy (the same material that was used for the metal electrodes in the fuel cell). That way I could purchase one item that would serve two purposes and thus save on costs. Stainless steel would also be a good choice – but it should be good quality stainless. I have found that nickel has many qualities that I consider superior, so I prefer that metal for use in an electrolyzer. Although a thicker gauge may have been better, it was much more convenient to be able to cut, drill and bend the material easily with common tools.

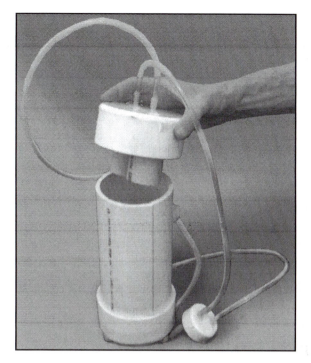

The P38 electrolyzer

Preparing the parts for assembly

PVC pipe can be cut with a hacksaw, bandsaw, or on a table saw or miter saw, and you can use a regular wood saw blade. I prepared the top cap by drilling two holes in it to receive the 1/4" barbed hose connectors (see templates). Inside each cap there was a little nubbin in the center from the cap's casting mold. This had to be removed in order to seat the gas collecting tubes without interference. I removed the nubbin with an abrasive tipped bit that fit into my hand drill, and then set the barb connectors with epoxy to make sure that the seal was tight.

I cut two pieces of the 1½" pipe to 3½" length for the gas collectors and glued them with epoxy to the top cap. I epoxied the inside rim and the outside rim very thoroughly to be sure that there were absolutely no gaps. It is important for these tubes to be sealed well so that gas cannot escape from the collector tubes. I let the epoxy harden for 24 hrs. or so and then

I tested it by filling the collector tubes with water to see if anything leaked out. I discuss this process further below.

Bottom cap

I also removed the bump from the inside center of the bottom cap and drilled five holes (see template). Four $7/64$" screw holes were drilled to accept the binding post and holding screws for the electrodes. A $1/4$" hole was drilled in the center to accept a 90° angled barbed hose connector for the height gauge filling tube. One flared end was cut off the connector and the connector set in place and epoxied well so that it would not leak.

Inside view of the electrolyzer top

Looking down into the electrolyzer, with the electrodes in place. The height gauge tube is visible at the upper left. It is connected to the black barb splicer, visible in the center.

Height gauge

The height gauge filling tube serves two purposes. It allows you to see what the electrolyte water level is in the tank, and also serves as a filling tube to replenish the water used. As the process of electrolysis proceeds, the water is used up as it is dissociated into hydrogen and oxygen. The potassium hydroxide is not used up in the process. When the water-electrolyte reaches a certain level, it must be replenished with distilled water.

90° barb splicer

Maintaining the electrolyte level

For the P38, I use the bottom edge of the top cap as a marker. I usually keep the water-electrolyte level a tiny bit higher than the bottom rim of the top cap. The inside gas collectors extend beyond the bottom rim of the top cap. If the water-electrolyte level were to fall below the bottoms of the gas collectors, then gas would diffuse out into the top of the tank and would mix.

Mixing is not good because the fuel cells will not run on the mixture, not to mention that the mixture is explosive. Thus it is important always to check to see that the water is replenished to the correct level.

The P38 was designed to produce just enough gas to run a small bank of fuel cells. It does not generate copious amounts of hydrogen. The collection tubes can only store small amounts of the gas and the surface area of the electrodes is small, thus limiting the amount of gas generated – all of which makes it safer to operate.

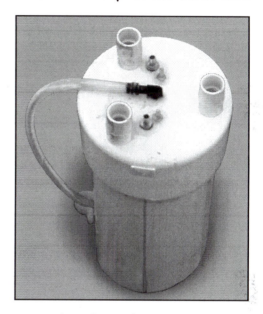

The electrolyzer upside down, without the top cap.

Increasing current to the electrolyzer

With more current going into the electrolyzer, the hydrogen and oxygen in the water would be disassociated faster. One could connect a rheostat on the electrolyzer's positive input terminal and vary the supply of current available for the process. This would increase or decrease the gas flow rate as desired. An ammeter is useful to monitor how much gas is made with varying amperage and to note the correlation. The larger the surface area of the electrodes, and the more current supplied, the more gas will be generated. If you design an electrolyzer you can experiment with these different parameters.

A recycled rheostat

Sealing the electrolyzer

It is very important that the epoxy seals are good when installing the water level-electrolyte barbed hose connector and the screws from the binding post terminal and other holding screws for the electrodes. They are on the bottom cap and if the electrolyzer leaks from these points, the conductive water-electrolyte will short out the positive and negative terminals, which are also located on the bottom. So, it's important to have tight fits between the barbed hose connector, binding post terminal, holding screws and their respective holes on the bottom cap.

Silicone cement vs. epoxy

For the screw holes for the binding post screws and holding screws, I used silicone cement in case I had to change electrodes. The silicone is much easier to remove than epoxy. When it dries, it remains rubbery and soft yet provides an excellent seal.

The first prototype of the P38 had cheap steel for the electrodes. After just a short time of use the electrodes were coated with sludge and I quickly understood that this cheaper type of steel was not what I needed. The cheap steel was replaced with nickel and I was very glad that I didn't epoxy the electrode screws.

Preparing screws and nuts

To prepare the electrode screws I coated the threads with silicone glue and inserted the screws in the holes, then put the nuts on. Before I fully inserted the threaded screws I put silicone cement on the surfaces that would touch: between the nut and the pipe cap and between the screw head and pipe cap. After securing these nuts tightly, I then covered the nuts and screws inside and outside the cap with more silicon cement to seal from any possible leakage. The point is to make sure there is a good seal.

Epoxy the pipe to the bottom cap

After inserting and seating those components, I spread an epoxy layer on the inside rim of the cap and the outside rim of the pipe for the full length that both would mate. I was liberal with the epoxy in order to assure there would be no gaps for leakage. I then inserted the pipe into the bottom cap and with a long stick (wooden shish kebab stick) dripped epoxy and rubbed it along the inside bottom seam where the cap and pipe came together. I took my time with this and used a lot of epoxy to make sure that I was totally covering and sealing.

Dry the seals and test

When I finished this operation I, epoxied the outside wall of the pipe where it meets the cap around the rim and again put lots of epoxy to avoid leaks. I then glued 3 small PVC couplers to the bottom side of the cap for feet. I let this dry for 24 hrs. and then filled the tank with water and let it sit for a while to test for any leaks. I had a few and put more epoxy in spots I did not cover well before. Be sure that your surfaces are dry if you re-epoxy. After the tank is emptied it will be wet, so it has to be dried off before applying the rest of the epoxy. There is an epoxy that can be applied to wet or damp surfaces but I have not tried that and cannot vouch for it. Of course I could have avoided going over this again had I done a better job in the beginning.

Aligning the top and bottom caps

On the outside of the 4½" PVC tank pipe I have a line going vertically from the top of the pipe to the bottom. This is a marking line that is put on when the pipe is manufactured. I use this line to properly align the top and bottom caps.

The bottom and top cap templates have a point marked for aligning the top and bottom to each other. From that point I marked straight up the inside of each cap over the top lip of the cap and down the outside, far enough to see easily. As I glued the bottom cap on, I lined up the line on the cap with the line on the pipe. As I pushed the top cap into place I made sure the line on the cap stayed on course. This ensures that the gas collectors are in line with the electrodes and the gas will be going where it is supposed to go.

Wiring it up

I used "zip cord" – 10 gauge stranded red and black wire – to connect the electrolyzer to the power supply. At the electrolyzer end of the two (red and black) wires I crimped loop connectors, to connect to the red and black terminal posts (red to red, and black to black).

I have gotten into the habit of using the standard color coded wires and terminal posts for projects as it makes things neater and safer. It quickly shows you what wires to hook up where, and prevents a lot of wiring mistakes.

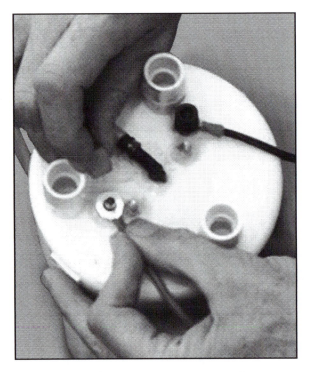

Attaching connections. The electrolyzer is upside down.

Filling the electrolyzer

When I finished testing for leaks and everything else checked out, I filled the electrolyzer using a squeeze bulb siphon, which is a safer way to put in the KOH solution than pouring, which can cause splashing. I filled it up to a point above the inner gas collection tubes. I had to guesstimate this because when the cap is put on it will displace some of the solution. It is better to under-fill it and then fill it to the desired level by adding more electrolyte via the water level-feed tube with the funnel.

Connecting to power

In my lab I have a bank of six 6 volt Trojan deep cycle batteries that I use for test purposes. I hooked up the electrolyzer to one of these batteries – the red wire to the positive terminal of the battery and the black wire to the negative terminal of the battery. Voila – it worked beautifully! Both electrodes began gassing immediately and I noticed much more gas coming off the hydrogen electrode (negative – black color) than the oxygen (positive – red color) electrode, as it should be.

I disconnected the battery from the leads, and aligned and pushed the electrolyzer top down making sure that the cap stayed in alignment as it seated all the way.

During the first test operation I tested the current draw of the electrolyzer by connecting a multimeter in series on the positive wire. I tested the electrolyzer with a home built 32 watt solar panel as a power source and then a 6 volt Trojan battery as a power source. The current draw from the panel was around 2 amps which was the panel's rated output. The draw from the Trojan was around 5 amps. Both power sources were more than adequate for powering the electrolyzer, although gas production from the battery was greater due to the higher current available.

The current draw of an electrolyzer depends on certain variables. These are: surface area of the electrodes; how far the electrodes are away from each other; and the strength of the concentration of the electrolyte. There are other factors, but these are the most important.

I specifically designed this electrolyzer to be a low flow, low pressure device as I was not particularly interested in producing a quantity of hydrogen for storage, and wanted to have a unit which would be rather portable and would be relatively safe to use but at the same time supply enough gas to power a number of fuel cells.

Scrubber

As the gas exits the electrolyzer chamber it will contain some caustic vapor and very small particulates that can affect the performance of your fuel cell over time. In order to scrub out these undesirable elements, the gas is run through a solution of natural vinegar or water, or through both, to soak up the impurities.

Vinegar

Vinegar scrubs the potassium hydroxide by neutralizing it and picking up any particulates. Water also soaks up the caustic vapors and picks up particulates. This gives a cleaner gas entering the fuel cell and will add to the longevity of the fuel cell's operation.

The vinegar should have no sulphur additives. Even if it says "natural" on it, check the fine print. Pure vinegar is available from health food stores and food co-ops, and many regular supermarkets now carry vinegar that has no additives. Either white or red vinegar is fine.

Red vinegar will turn pale when it has been spent which is a nice visual indicator. However red vinegar is more expensive than white. The white is fine if you just remember to change it now and then. I use red wine vinegar that is 5%. When operating my electrolyzer I use vinegar in one scrubber and distilled water in another.

Building a scrubber

The scrubber consists of 1½" PVC pipe that is cut to about 9¼" length with two PVC end caps to fit the pipe. The bottom cap should be epoxied on well so that there are no leaks. Drill two holes in the top cap, one hole to accept the tube from electrolyzer and the other hole for the tube to the gas distribution system, or to the cell. One hole should accommodate a ¼" barbed hose connector and the other, a ⅛" barbed hose connector, unless you wish to use other size tubing.

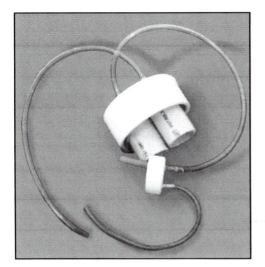

The electrolyzer top cap, connected to the scrubber top cap.

Cut the flared end off one side of each of the ¼" and ⅛" barbed hose connectors and push them through the holes drilled for them. Extend the end of the ¼" barb splicer through the cap so that on the other side there is just enough jutting out to easily attach a small tube on that side. The ⅛" barbed hose connector can be flush with the inside.

Epoxy the barbed hose connectors. Apply epoxy to both sides of the cap where the hose connectors jut through and use plenty of epoxy to avoid leaks. Let it dry, test for leaks. Connect a small piece of hose, about 1" long, to the portion of the barbed hose connector inside the cap. This tube will extend down into the vinegar when the scrubber is filled. Fill the scrubber with vinegar, put on the top cap and make sure that the tube on the inside is below the liquid level of the vinegar. You would have to guesstimate this if you use regular PVC pipe rather than a clear PVC or other such clear plastic.

The next step is to connect the gas tube from the electrolyzer to the ¼" barb splicer and a tube to the ⅛" barb splicer that connects to your fuel cell.

When the electrolyzer is turned on, it will take a little while for the gas to begin to bubble through the scrubber. It will go through the tube, bubble into the vinegar, then rise to the top and exit out the other hole towards the fuel cell. Distinctive "bloops" can be heard as this occurs.

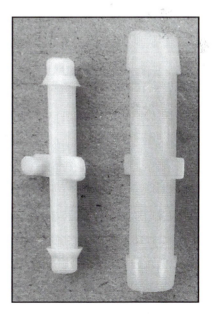

Barb splicers

Inside tube length

This inside hose could be a little longer, perhaps 1½", but there should not be too much water pressure on the exiting gas from the electrolyzer as that would tend to back up in the electrolyzer. The tube in the scrubber bubbler should be about ½" or so below the liquid level of the vinegar. The deeper you insert the tube in the vinegar, the more back pressure. This could stop the gas flow and force it out from the rim of the cover on the electrolyzer (which is not sealed and not designed to be operated under much pressure).

Also, with back pressure, the gases can mix in the electrolyzer if they are forced out of their collection tubes back into the top of the electrolyzer. In this case, the electrolyzer top must be removed to release this mix. A fuel cell cannot run on mixed gases.

Experiment and see what works for you. A double scrubber can be made by connecting up to two PVC pipe scrubbers, one filled with vinegar and the next one with distilled water, as long as the inside hose doesn't go too far into the liquids – again it should be about ½". The liquids should be changed occasionally to ensure appropriate scrubbing action.

Transparent scrubbers

Clear PVC or other see-through plastic tubing is great for watching the bubbling action and the change of color if you use red vinegar; and, if you use distilled water, the water gets a yellowish tinge after a while. I used common PVC as it was readily available at lower cost. McMaster-Carr and other industrial suppliers have clear PVC and other plastic see-through pipe that would work for the scrubber and/or electrolyzer. Electrolyzer and scrubber systems should be checked to confirm that they are working properly before they are connected to fuel cells. When not in operation, bubblers and any connection from the electrolyzer directly to a fuel cell should be disconnected. This will prevent liquids from being sucked into the cell from pressure. The electrolyzer should always be disconnected from its power source when not in use.

Making an Electrolyte

Making your own electrolyte is easy if you have access to hardwood ashes. I burn wood as our main heat source so it would be silly for me not to take advantage of all that free caustic potash (potassium hydroxide) that I can make from the wood ash.

Tools needed

A KOH (potassium hydroxide) solution is easily made and only requires two five gallon plastic buckets with lids, a drill to make holes in one of the buckets, some small clean stone (about bean sized), two or three clean rocks (more or less the size of 2 fists), 5 gallons of distilled water, and a bulb-type electrolyte siphon available at any hardware, discount, or auto store. You also need to wear rubber gloves, a respirator and eye protection, as well as clothing that leaves no skin exposed.

The filter bucket

Turn one of the buckets upside down and drill a bunch of 9/64" (more or less) holes all over the bottom of the bucket. Don't skimp on them – you want as many small drain holes as possible. Get some pea-sized stones to cover the bottom of the bucket with the holes in it with a 1½" to 2" layer of the small stones. This will be the filter.

Drill a bunch of holes in the bottom of one plastic 5 gallon bucket to make the filter bucket.

The drip bucket

The other bucket is the drip bucket. Place two or three large stones in the bottom of this bucket to prop up the

87

Drip bucket, left, and filter bucket

filter bucket when it is placed inside the drip bucket. The stones will have spaces between them for the KOH solution that drips down from the ash bucket. They also provide some height to the filter bucket so that the liquid being collected will not soak back into the ashes, and their weight gives the two buckets more stability so that they will not easily tip over.

Add ashes and water

Fill the filter bucket with wood ash to more than ¾ full or about 2" from the top. Place it in the drip bucket, resting on the rocks, and begin pouring 2½ gallons of distilled water into the ashes. You will not be able to pour all of it in at once. The ashes will initially take around 2 gallons, then wait until the water drains down a bit to pour in the last half gallon.

Add the ashes to the filter bucket. Filter bucket is set into the drip bucket

Add the water to the ashes.

Wait 3 days to a week

When all the water is poured into the ashes, put a lid on the ash bucket and let it sit for at least 3 days or up to a week, somewhere out of reach of children. Don't be fooled by the fact that most of the water drains through on the first day. I have found that there is quite a difference in the quality/strength of the KOH if the ashes and water sit and drip for 3 days to a week.

Collect the KOH solution

Wear rubber gloves, eye protection, and protective clothing. To collect the KOH solution, remove the ash bucket and put it aside. Use the bulb siphon to suck up the KOH from the drip bucket and squeeze it out into the empty poly gallon jugs that the distilled water came in. There will be sediment on the bottom of the bucket, so try to avoid sucking that up with the solution. If some sediment is sucked up, it will settle to the bottom of the poly jugs. No sediment should get into the electrolyzer.

There will only be a half gallon, or a little less, of liquid in the drip bucket. The strength of the dark yellowish to brown colored electrolyte solution can be measured with a hydrometer (available at auto supply stores, etc.).

Put the lid on the jug, mark it and store it in a safe place (out of reach of children). You then repeat the whole process with another bucket of ashes and 2½ more gallons of distilled water to fill up the rest of the jug.

This should be plenty of electrolyte to fill the electrolyzer. The solution will be somewhere near 1.1 strength. For a stronger solution, the lye can be simmered over a heat source (don't do this in your kitchen and never leave it unattended!), and tested with a hydrometer frequently during the "cooking" to see if it has reached the desired strength. The solu-

Lye in the drip bucket

tion from the bucket will be more than adequate for the electrolyzer as is and you do not have to simmer it down, unless you want to make a stronger solution.

Though the higher concentration is not necessary, gas production is more copious and efficient with a concentration from around 23% to 30%. Resistance in the solution at a concentration of about 21% is about 1.96 ohms per centimeter cube. At a concentration of around 25%, the resistance is about 1.85 ohms. At a concentration of about 29 the resistance is about 1.84 ohms. At a concentration of about 33% the

The author transferring lye from the drip bucket to a storage jug.

Testing the lye in the drip bucket with a hydrometer

resistance is about 1.91 ohms. A stronger solution beyond around 29% is not useful as the idea is to make a solution with a concentration that provides the least resistance. A solution of about 29.4% will give you the least resistance, which would be 1.84 ohms per centimeter cube.

The solution you draw off will probably be much weaker – perhaps around 12 to 13% solution, which would have a resistance somewhere around 2.60 ohms per centimeter cube.

The KOH table below correlates specific gravity, percent of KOH and pounds per gallon needed to produce your desired results. For those who purchase KOH at a chemical supply house, this shows how much you'll need.

Specific Gravity	Percent KOH	Lbs. per Gallon	Specific Gravity	Percent KOH	Lbs. per Gallon
1.0083	1	0.0841	1.1493	16	1.535
1.0175	2	0.1698	1.159	17	1.644
1.0267	3	0.257	1.1688	18	1.756
1.0359	4	0.3458	1.1786	19	1.869
1.0452	5	0.4361	1.1884	20	1.983
1.0544	6	0.528	1.1984	21	2.1
1.0637	7	0.6214	1.2083	22	2.218
1.073	8	0.7164	1.2184	23	2.339
1.0824	9	0.813	1.2285	24	2.461
1.0918	10	0.9111	1.2387	25	2.584
1.1013	11	1.011	1.2489	26	2.71
1.1108	12	1.112	1.2592	27	2.837
1.1203	13	1.215	1.2695	28	2.966
1.1299	14	1.32	1.28	29	3.098
1.1396	15	1.427	1.2905	30	3.231

For those who make their own solutions, the KOH table will give you an idea, via your hydrometer readings, how much KOH you are getting per ash bucket. Since there are a lot of other constituents in wood ashes that will affect your readings, it will not be totally accurate, but it will give you a ballpark look and is good enough for our purposes.

As mentioned previously, KOH is inexpensive, but if you don't have a source where you can pick it up yourself, the hazmat fee plus shipping costs make it pricey. Do not work with lye if you do not wear safety goggles or face shield and proper protective clothing. A little splash in the eye for instance could cause blindness for life! I repeat – respect this stuff and do not work without proper protective gear. Read the MSDS and follow all safety, and disposal directions.

Building the L78 Soft Graphite Single Slice Fuel Cell

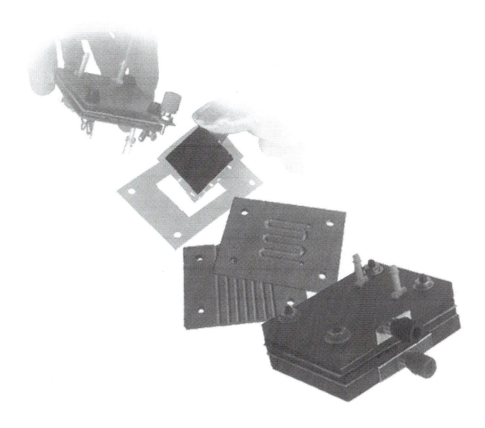

Tools

Inexpensive plastic art paintbrush. The end of the brush handle is used to make the gas flow grooves in the graphite foil. Any round tipped stick that is about 1/8" in diameter will do the job. The brush end of the paintbrush is good to brush away the graphite flakes as the grooves are made.

Knife or other cutting instrument such as utility razor knife, sturdy razor blade with backing, Exacto knife or Exacto medium to fine blade saw. I found that an old table knife I had worked best for cutting the graphite foil. Test some pieces of graphite foil with your cutting tool of choice. I also use a smooth metal Exacto knife handle, to roll over the graphite surface to smooth it as I worked.

Saw. I use an Exacto saw with a medium to fine blade to cut the circuit board. A hack saw or many other types of saws would also work.

Q-tips. To clean and smooth the surfaces of the graphite while making the grooves.

Hand drill. Either hand or power

Drill bits. Same as for the K18 fuel cell, 7/64, 9/64, 7/32.

Hard working surface. I use a polished stone surface, but any-thing – metal, tile, Formica, etc. – that is hard, smooth and doesn't give will work.

Pliers.

File. If double clad circuit board is used, a small portion of the out-side facing plate where the binding post is attached will have to be filed or sanded off.

Scissors or razor blade, Exacto knife, etc., to cut Mylar and rub-ber gasket material

Small diameter 7/32" hole punch. Trade size 8. Part #3424A18, McMaster-Carr

Materials needed

Dual or single sided copper clad circuit board. Can be purchased from any electronics supplier. Radio Shack has a dual sided PC Board 4½ x 6³/₈", part #276-1499. One piece is sufficient for making the two end plates.

Graphite foil. One 12"x 12" graphite sheet gasket material, McMaster-Carr part #95715K63

Thin gauge silicone rubber sheet, .020" thick. One 12"x 12", 35A durometer, McMaster-Carr, part #86435 K45.

Mylar sheet. Comes free as carrier sheet for silicone rubber sheet when you order from McMaster-Carr

Insulated binding posts. Two, available from electronics suppliers. Radio Shack Part #274-661.

Screw fasteners, nuts, washers, wing nuts, or regular nuts. Available at local hardware store. Four- 10-24 x 1" socket head cap screws. Four 10-24 wing nuts or regular nuts. Four insulating fiber washers to fit screws.

Membrane electrode assembly (MEA). See MEA Suppliers, page 173.

Liquid electrical tape, liquid rubber dip or shrink tube. Liquid electrical tape is available from local electronics stores, mail order suppliers, hardware and electrical supply stores and auto stores. Liquid rubber can be found at discount and hardware stores and is used to coat tool handles etc. Shrink tube can be found at the above sources also. Shrink tube is used to insulate connectors, etc.

Barb splicer hose connectors, 1/8" x 1/8". Two. Can be found at hardware stores, or Small Parts, Inc.

Silicone rubber adhesive sealant. Local hardware stores

Tinnit (optional) Bright tin electroless plate will help to preserve copper surface from oxidation. Can be purchased at electronic stores, or from All Electronics, part # ER-18 (check as this part # may change).

Really Simple Tools Do the Job

Stick technology

This version of a PEM fuel cell is very easy to make. I developed it so that it could be made without using any power equipment whatsoever – what I like to call "stick technology." The only tools needed are a hand drill (either power or not), a dull knife, scissors or Exacto knife, magic marker, etc.; and the "stick": a plastic artist's brush, or chopsticks, or a wooden dowel with a rounded end. The stick is used to form the flow fields in the graphite foil plates.

The components are easily available and consist of copper clad circuit board, graphite foil, silicone rubber gasket, binding posts, and of course, Mylar with inserted MEA (membrane electrode assembly).

Make it on your kitchen table

The circuit board is both electrode (current collector plate) and end plate, so there is no need to cut out metal collector plates. This fuel cell is much thinner, does not use silicone rubber spacers and has only one silicone rubber gasket. It can literally be fabricated on your kitchen table. The graphite foil is available from McMaster-Carr and sold as "graphite gasketing sheets." It is very easy to work with and requires a minimum of skill.

Copy the templates

The templates for this project can be found on pages 189-193. You may find it helpful at this point to refer to them as you read the directions. If you make copies, check the printed templates with a ruler to be sure that they have printed out at the correct size. Measurements are marked on the templates.

Endplate/current collectors

The end plate current collectors (circuit board) can be purchased at any electronics supplier, i.e. Radio Shack. The "double sided" and "single sided" refers to the copper layers. Double-sided has copper on both sides of the circuit board; single has copper on only one side. Either double or single sided board will work, though it is probably best to use single sided board.

To function, the fuel cell only needs a copper layer facing each graphite collector flow field plate. The copper provides a conductive surface that connects with a graphite foil plate and conveys the electricity to the binding posts, just as the nickel did in the K18 fuel cell. If you want a shiny look for the completed fuel cell, use the double sided board as you see in the photo, but you will have to file or sandpaper off the copper on the outer side at the binding post tabs so that current is not conveyed to the rest of the outside of the plate.

A small Exacto saw easily cuts the circuit board, but you can use anything you wish. When hand cutting materials it is important to be able to control the cutting process as much as possible, and I found the small Exacto saw to be just right for this purpose.

Mark the end plates

Print and cut out the templates for the end plates, place them on the board and draw the outline of the pieces to be cut. Mark the gas holes and the binding post holes. To mark the holes, I simply took a pointed dental tool and put it at the center of the cross-hairs on the template and pushed downward. This leaves an indentation for exact positioning of the drill. Any sharp pointed object such as an ice-pick, etc., will work. Do not pound an indentation in, as this

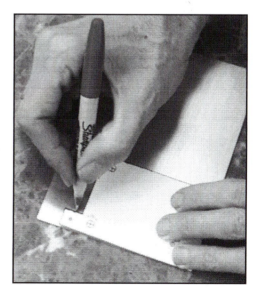

Use the templates to mark the circuit board

might cause the surface copper plate to fracture. Simply push and twist hard to make a good indentation that you can use as a guide for the drill and make sure that you are centered on the cross hairs.

Cut and drill the end plates

Next, cut the circuit board and drill the holes. Be as accurate as possible when you cut the plates. Take your time and figure out the best way to proceed that will give you the most accurate size cut. Go slow with your sawing. To drill, move the pointed drill bit gently along the surface and let it drop into the indentation, then begin to drill. The hydrogen side

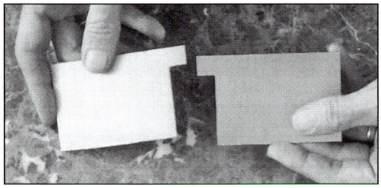

Above, cutting the end plates from the circuit board

Left, the two endplates, ready to drill

plate will have two holes for the gas ports. Be as accurate as possible when drilling these holes as they have to line up very well with the gas port holes in the graphite plate.

To complete preparation of the end plates, file or sand off the tabs on the outside of the plate so that current is not conveyed from the binding post screw to the rest of the outside of the plate. This is not necessary if single sided circuit board is used.

Soft graphite

Graphite foil gasket material is made by taking graphite cloth and pressing it into a foil. Layers of this foil are then pressed together to form the desired thickness. No binders are used in this process so it is a mechanical bond. This produces a unique material that is layered, and the layers can be separated easily, if so desired. You can prod off various layers from the edge of your practice piece to see this yourself.

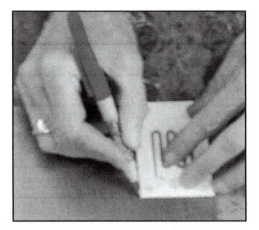

Using a template to mark the graphite foil

Practice with the material

Before preparing the final graphite parts for assembly, cut off a piece of the graphite foil and practice cutting and drilling this material. See what tools work best for you. It's very easy to work, once you get a feel for it.

Cutting the soft graphite

The next step is to cut the graphite foil to size. Print and cut out the template, lay it on the graphite and outline with a fine pointed felt tip marking pen (like a Sharpie). You can use any cutting tool that gives an accurate cut on the graphite foil. I used a table knife. If you use a very sharp cutting blade such as a utility razor knife, go very slowly. With this material it

Two plates cut from graphite foil.

is easy to lose control if you try to cut in one fast swoop. Take your time to cut as accurately as possible.

Working with soft graphite

When cutting the graphite foil, work on a hard, smooth surface that has no give, and when drilling, use an underlay of stiff plastic or fiberboard that does not have much surface give.

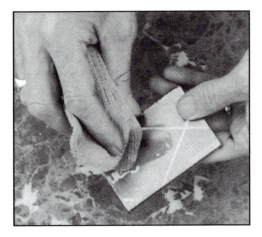

Wipe off the printing on the foil back

After cutting the plates, remove the printing on the plates' backs by wiping them off with a rag and denatured solvent alcohol. Make sure that all paint residue from the printing is removed, then wash the plates in distilled water to remove all traces of solvent and residue. These impurities can interfere with conductivity.

Drilling holes in soft graphite

When you drill holes in the graphite foil a small mound appears around the edges of your hole. The graphite is so soft that it is pushed up and out during the drilling process. These mounds have to be leveled off when finishing the plate. This can be done in several different ways. I moved a stiff

*Smoothing the edges of the graphite plate
with a smooth metal tool handle*

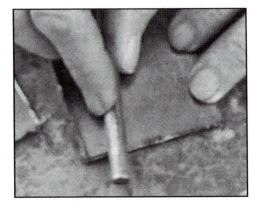
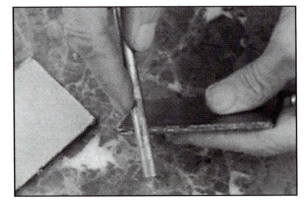

razor blade more or less flush along the surface and cut the protrusions off. Then I rolled the smooth handle of my Exacto knife over the surface of the hole – a bit like using a rolling pin to even out dough. Any smooth metal rod will work.

The graphite will tend to compress back into the hole a bit, making the hole smaller when you do these operations. To bring the hole back to size, take the drill bit out of the drill and twist it by hand back into the hole, and then twist it out. Repeat the cutting around the edges of the hole with the razor blade and rolling it again. Keep repeating these operations until the holes are the appropriate size, and the surface is smooth and without any protrusions. The surface should be as smooth and flat as possible, though it doesn't have to be perfect.

During these operations, be careful not to flake the graphite foil with your pushing and prodding. The foil has a tendency to separate in layers. You will have to gently push some of the material around the holes gently in from the surface to the inside of the hole. Don't press too hard but just enough to maintain the integrity of the layers. This material tends to crack and flake, but as you work with it you will find that it is easy to shape and form once you are familiar with it.

For all of these operations you can use anything you can think of to smooth out the surfaces. I used both the Exacto knife handle and the end of the paint brush to smooth out the surfaces around the holes, as well as my fingers.

Mark the flow fields

Using the templates, mark the flow fields by cutting a piece of carbon paper the same size as the template. Lay the carbon paper down on the graphite with the template on top, and transfer the flow field markings to the graphite with a pencil, etc. Another method is to simply lay the template on top of the graphite piece and with some sort of stick with a rounded tip, about $1/8$" diameter, gently press along the lines of the flow field to make a light but distinct indentation. Move up and down the

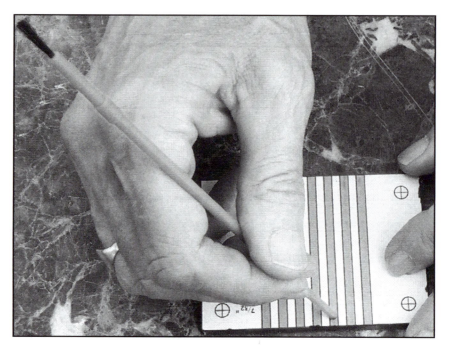

Marking the flow fields on the graphite plates

lines and press a little harder each time, just enough so that when you take the template off, you have guideline grooves to follow to make the final deeper grooves. After the flow field markings are transferred from the templates to the graphite foil, remove the template and continue to deepen the impressions.

Making the flow field grooves

I used the end of an inexpensive artist's brush to make the grooves, but chopsticks, etc., will also work (use whatever is available, whether wood, metal, plastic, etc.) The end or tip of the brush handle or stick should be the width of the groove, about $1/8$". Hold the brush or stick as vertical as possible while making the impressions. This will give your grooves a uniform width. Trace through the grooves gently, applying pressure to deepen the whole groove a little at a time. Do not try to make the full depth at one pass. Each time you go over the pattern in the plate you will impress deeper and deeper. After each pass, brush out the flaky carbon.

Continue this process until you have reached the appropriate depth. The foil is only $1/8$" thick, so be careful about the integrity of the material as you impress deeper and deeper. You do not want to weaken and break through the material, but at the same time the grooves should be as deep as is practical, at least $1/16$" and more, if possible.

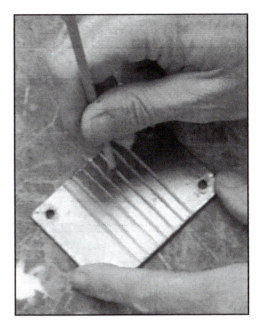

Making the flow fields with a paint brush handle.

When making the grooves, work on a hard surface. The pressure tends to push the graphite and can create an outline bulge on the back. Turn the plate over as you work, and you will notice the outline on the back bulging out from the pressure you have applied. A soft work surface will make the indentation more pronounced, which is not good because the bulging should be kept to a minimum. I use a slab of marble for a work surface.

When pressing the vertical grooves for the oxygen side plate make sure that at the edge of the plate, the depth of the grooves is the same depth as they are in the interior of the plate, or a little deeper to assure sufficient air flow. When you impress the grooves near the edge, press in a one-way motion, from the inside toward the edge. This will helps to avoid flaking the edge surface more than necessary which could compromise the structure of the groove.

Roll, cut and rub

Pressing the grooves will build up a ridge on either side of each groove Remove this carefully with a razor blade and burnishing techniques just as you did for the drill holes. I used my Exacto knife handle to roll over the whole piece, perpendicular to the grooves, applying pressure with the palm of my hand. Then do a little shaving, a little spot burnishing and then some more rolling, repeating the process until you have a fairly smooth surface.

As you burnish and roll, more material will fall back into the groove. Reapply your stick to the groove, moving it along to clear it out. Then re-burnish and roll and reapply the stick and so on until the surface is smooth. Q-tips also work well to smooth and clear out debris from the grooves in this process.

Soft graphite is forgiving

As always, try out these techniques on a practice piece to get a feel for it. I chose graphite foil for the L78 because it is easy to work with, requires minimal tools and is fairly forgiving if you make mistakes. The grooves and surfaces do not need to be picture perfect. The foil itself acts as a gasket and when the fuel cell is assembled, tightening the fasteners will compress everything down to a good seal if your mistakes are not too outrageous. This fuel cell would also be a fairly easy classroom project.

Easy construction and high performance

Do not be misled by the simplicity of this fuel cell or the fact that stick technology is used. If made properly, this cell performs just as well as the hard graphite fuel cells or any other commercial fuel cell available.

Aligning the holes

For this fuel cell, the barbed hose connector goes through both the circuit board end plate and graphite collector flow field plate. In the hard graphite fuel cell, the barbed hose connector only went into the end plate. This makes it even more important that the gas port holes line up well between the circuit board plate and graphite plate.

Template, graphite plate and end plate

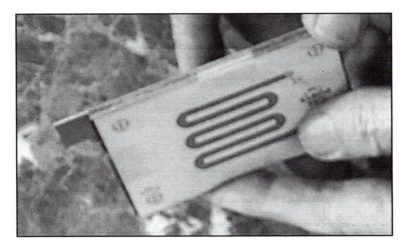

Align the holes in the hydrogen side end plate and hydrogen side graphite plate by taping them together with the template, then drill the holes through both layers at the same time.

The less exacting your cutting and drilling, the more the alignment will be off within the fuel cell. With this design, you can drill the hydrogen side circuit board plate and hydrogen side graphite flow field plate at the same time to get a more exact alignment.

If the holes are a little off, it will not make too much difference, as long as the gas flow is not blocked. I have constructed some cells with slightly misaligned holes and they work fine as long as the gas holes are not totally blocked from each other.

The hose connectors

To prepare the 1/8" barbed splicer hose connector, cut off one of the ends just behind the flare and insert the connector through both the end plate and graphite plate. Note how much excess sticks out, and mark it with a razor blade. Remove the barb splicer and then cut to the mark.

Prepare a very small amount of epoxy and coat the outside of the barb splicer at the points where it will be contacting the plates, then reinsert the barb splicer. Allow enough protrusion of the barb splicer on the outside of the circuit board end plate to connect well with your hose.

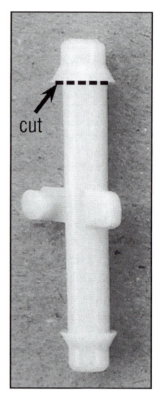

Cut one end off the barbed splicer to insert

The barb splicers that I purchased had a tab in the very middle which I used as stop gap against the outside of the end plate. If you don't have this stop gap, make sure to leave enough of the connector protruding on the outside for a good mating with the tube.

On this fuel cell, the hose connection is a bit more delicate than with the heavier hard graphite fuel cell, so it's good to put a length of hose on the fuel cell's connector for the gas entry port and leave it there permanently. Connect and disconnect to the gas source at the other end of the tube to avoid stressing the fuel cell's barb splicer connection. You can always re-epoxy the connector, but it's better to avoid the problem.

Prepare the fasteners

The fasteners can be insulated by giving the screws a couple of light coatings of something like Star Brite liquid electrical tape. This comes in a small can with a brush attached to the lid for application. It forms a rubbery surface and is a good electrical insulator. A dip-it liquid rubber compound (for dipping the handles of tools) will also work. Shrink tubing, as used for the hard graphite fuel cell**, is good, too. Just use a smaller length of shrink tubing as this fuel cell is a lot thinner than the hard graphite cell, so measure accordingly. Instead of regular nuts, I used wing nuts so that I could hand tighten the plates easily, and I used fiber washers rather than steel as on the hard graphite cell. Be sure to put the wash-

ers for the screw heads on before you shrink the tubing and/or coat with liquid electrical tape.

Prepare the MEA

Cut out and punch the Mylar surrounds according to the template. Glue the MEA into its Mylar surround as in the instructions for the hard graphite fuel cell and wait for everything to dry properly before putting the cell together.

Assemble the fuel cell

When all of the components are ready, simply put the pieces together and attach your binding posts. (See the diagram on the next page for a blow-up view of the cell layers.) Hand tighten the screws until they are snug, but don't overdo it. To test the cell, connect it to your electrolyzer and observe the reading on your voltmeter. If you are not getting a reading, hand tighten the wing nuts a little more. Also, if you are getting a reading that seems low, hand tighten a little more to see if the voltage increases. In some cases you my have to loosen a bit if it is too tight. By going

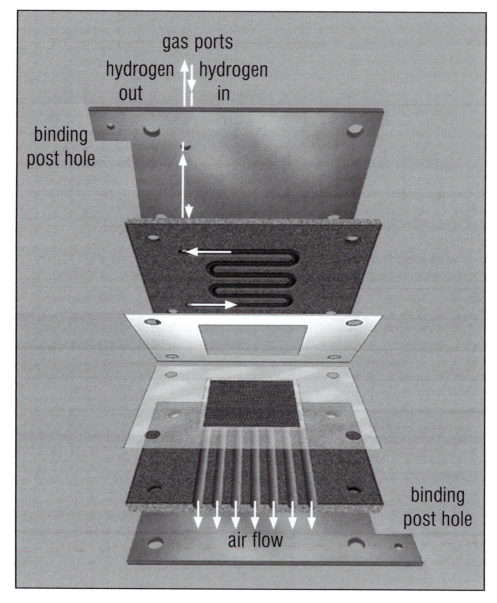

gas ports

hydrogen out   hydrogen in

binding post hole

binding post hole

air flow

Layer view of the L78 soft graphite
single slice convection fuel cell

through this process you will get the right compression so the graphite plates are contacting the carbon cloth properly and also close any gas leaks if the plates are not tightened enough.

Designing Fuel Cells

Ideas for Designing Fuel Cells

There are many different ways to construct fuel cells with many different types of material. This book is intended to provide an introduction into the fabrication of fuel cells with a hands on approach that will give you the skills and experience to proceed further on your own. The concept and operation of a single slice fuel cell is the foundation of fuel cell technology. Working with these simple design and construction techniques will give you sufficient experience to add your own modifications with ease.

Fuel cell geometry

Consider adding slices to make a convection fuel cell stack configuration, or try a planar structure design. You can also create fuel cells in a cylindrical form – a tubular configuration with hydrogen piped in through the center of a tube – MEAs make up part of the tube wall, and oxygen/air is supplied to the MEA on the outside surface of the tube wall. The geometric possibilities are many. Imagine spheres, or geometric configurations like Buckminster Fuller's geodesic designs, and so on.

Surface geometry can have quite an impact on fuel cell operation, although it is not always readily apparent. For instance, in conventional stack design, heat and water can become a problem as the stacks become larger. More heat is generated with more slices, and water buildup can be a problem on one end, with a lack of moisture problem on another end due to excessive heat. All of these factors affect the efficiencies of the cell stack. Different shapes and configurations such as a planar design and/or the applications of new materials could solve these problems.

Bucky balls

At the nano level, experimenters are picking up on the geometry concept by using Bucky balls (Fullerenes) – geodesic structures in the form of carbon that are platinized to provide more surface area for the catalytic reaction. This reduces the amount of platinum needed and thus the cost. Also nano tubes have been experimented with for gas storage, and these hold promise for the future. At present, nano fuel cells, and biological fuel cells are being constructed, and this should be very interesting for future applications.

The application of quantum mechanics will probably make it possible to develop a fuel cell on a chip.

Ceramics in fuel cells

Researchers are exploring the use of clay (ceramics) for various components of a fuel cell. Clays make good electrolytes if doped properly, and they also are easy to mold for fuel cell flow field plates and other functions. Polymer clays may be good for membranes. Mond and Langer used a clay soaked in sulfuric acid for an electrolyte membrane back in the 1920s. It is fascinating how many new ideas are really based on old ideas.

There are many types of clay and rocks such as zeolites that have interesting characteristics for use in fuel cells. Certain types of pumice rock, for instance, would make a good gas diffusion device and could be made conductive. Conductive foam (which integrated circuits are packaged in to draw off static electricity) also makes good gas diffusion devices and is promising for fuel cell experiments.

Wooden fuel cells

At present there is a Canadian company researching the use of wood for flow field/conductor plates. Flow field grooves would be put into the wood either before or after carbonization. Carbonization of the wood produces a conductive carbon plate. Both wood in its natural form and in carbon form are easy to work, so these devices would be easy and extremely inexpensive to produce with the simplest of tools. I would presume that a good tight and heavy wood would be used. When you carbonize wood it loses about 70% of its weight. This makes sizing of the grooves before or after carbonization an important consideration, although the process of carbonization seems to retain the features of the original configuration of the wood pretty well. On the other hand, a partial carbonization of a wood block, for instance, just the surface area of one side, would suffice to create a conductive plate. This would seem to be easy to accomplish and would be very inexpensive to produce.

Carbonizing other materials

One could also carbonize any type of plant product and mill it down to a fine particulate (usually called "air float" in the charcoal industry), mix it with water and then compress it into conductive blocks with dies that will impress the flow fields. This is basically how charcoal drawing pencils and charcoal briquettes are made. Any type of material that can be carbonized can be used. I even experimented making a carbonized fabric from cotton. The carbonized fabric is conductive and, although delicate, does retain some of its flexibility and could possibly be used as a conductive gas diffusion layer.

Other natural materials for inspiration

Some plants such as cattails have root systems which store hydrogen, and experimentation with such natural materials might provide some interesting results. Consider the fact that nature has been making membranes for many years. Natural membranes are certain to have much to teach us about ion exchange and gas exchange. Plant leaves, for instance, will absorb certain gases and expel others. This is sophisticated technology that we are only beginning to understand.

If you think I am getting too far afield, consider that plants are now being used to generate hydrogen as a possible fuel source for the coming hydrogen economy. Nature has a plethora of models for developing new technologies. Drug companies routinely spend an exorbitant amount of money hiring biologists to rummage through remote rainforests to look for the next wonder drug that can be copied and synthesized in their laboratories. Other disciplines are now beginning to follow their lead.

"Biomimicry" for fuel cells

Don't limit your thinking or research – take an interdisciplinary approach to the design of fuel cells. Fuel cell technology has counterparts in nature. Look there and you may make some very exciting discoveries. For further reading about this kind of approach, I suggest *Biomimicry* by Janine M. Benyus, and chapter 19 of R.A. Ford's *Homemade Lightning*, which discusses scientific process and creativity.

Collect supplier catalogs

If you want to pursue fuel cell design, it is very helpful to acquire as many supplier catalogs as you can get your hands on, from chemical supplies to industrial supplies, and even from seemingly off-beat areas like art, photographic, craft, lapidary, fabric, etc. There are many tools and materials that you will accidently come upon that may be perfect for your experiments.

Learn from the materials

As you work with materials you will become acquainted with them and realize new and more efficient ways to work with them. For instance, when working with the graphite foil gasket material, I became aware that, in addition to being workable with "stick technology," with this material, parts could be stamped with the flow field design quite easily and quickly, and the material cut to size in one automated operation.

Some other design ideas that came up with while working with this material including using acid free paper with graphite rubbed into it to make it conductive. The paper could then be cut into a flow field pattern and backed with a thin piece of circuit board that had flow fields routed into it. This would then be backed by a thin piece of Mylar cemented to the back. This could be a very efficient, lightweight fuel cell, that would also be inexpensive and easy to construct.

Thinner and lighter designs

Electrical contacts and wires can be made by drawing lines and terminals with conductive epoxy for a more lightweight and low profile. Fuel cells can get by with extremely small gas flow fields and it should be possible for anyone in their workshop to produce almost credit card size thin fuel cells with a little experimenting. Researchers have also been finding that thinner membranes permit a greater power density, so the use of a Nafion type 112 membrane would appear to be more efficient than a 117 (which is thicker). There are also other manufacturers that produce membranes based on a slightly different chemistry.

Use easy to obtain materials and tools

When designing fuel cells, first and foremost, the materials that you use should be readily obtainable at a reasonable cost. Preparing the parts and assembling the cells should be easy and require only common tools.

For instance, to produce the K18, a bandsaw and drill press are used to prepare the graphite plates. In the L78, a 19 cent brush and a dull knife worth about a dollar, a small handsaw for $3 and a small hand drill for about $15 can be used. The same result was accomplished, but at substantially different cost for the tools and a substantial difference in the amount of work required to produce each type of cell.

End plate materials

Simple fuel cell design requires end plates with some rigidity. Many types of materials can be used at various thicknesses. Plastics are a good choice, but there also is a variety of natural materials such as natural stone – slate, soapstone (easy to work), etc. Other possibilities are phenolic sheet, various metals, composites or whatever you can come up with. The key is that the pressure exerted by the fasteners should squeeze the graphite plate evenly against the surface of the MEA. Any material that has some degree of stiffness and is not too elastic will suffice

Plastics in fuel cells

Plastic can be used for a number of fuel cell components and can be cast or molded. Molded epoxy plates are now being used in some fuel cells as flow field/conductive plates. The epoxy is mixed with graphite and molded in a form which has the flow field design.

Fuel cell parts can be fabricated easily with table-top digital 3D printers. You can design parts on a computer and print them out full form at the push of a button. This real time capacity can be very handy for experimental fuel cell designs.

General materials considerations

For any fuel cell part, consider different ways the part could be made to match your needs and resources. You could even use different materials for the same part in one fuel cell. For instance some materials will retain heat or moisture and some won't. For larger fuel cell stacks, factors like heat and moisture control can be critical for good performance.

Convection vs. oxygen-hydrogen

The two fuel cell projects detailed in this book up to this point were designed on the convection model. In other words, for the oxygen side supply, the oxygen is taken from the atmosphere, which is fine up to a point. However, you can feed oxygen from your electrolyzer to the fuel cell. Fuel cells with a pure oxygen input have efficiency rates up to 30% higher than those of a simple convection flow design. This is no small boost.

I chose a convection fuel cell model as a starting point because in larger stacks, convection cells are more trouble free. In larger oxygen-hydrogen stacks, water build-up can be a problem on the oxygen side, so getting rid of the water must be considered. As water forms on the oxygen side, it can accumulate and drown the cell. For a home based power supply, for instance, this consideration is very important. Also, it has been my experience that oxygen-hydrogen fuel cells run warmer than convection cells. Convection cells may be lower efficiency, but they are also lower maintenance.

Oxygen-hydrogen fuel cells

To build an oxygen-hydrogen fuel cell, simply make the oxygen side of the fuel cell the same as the hydrogen side; that is, the oxygen graphite plate would have a serpentine flow field instead of vertical, there would be oxygen side barbed hose connectors; and gas ports that go through the oxygen side components. There would be a rubber gasket on the oxygen side, the same as for the hydrogen side.

I have included two sets of templates for oxygen-hydrogen single slice fuel cells, one set based on the K18 hard graphite and the other set based on the L78 graphite foil cells. You may wish to print out the templates to refer to. The construction methods are pretty much the same as for the single slice convection cells. Just follow the templates and the appropriate layer view diagrams shown here.

Layer view of soft graphite single slice oxygen-hydrogen fuel cell. See construction directions for the L78, pages 93-108.

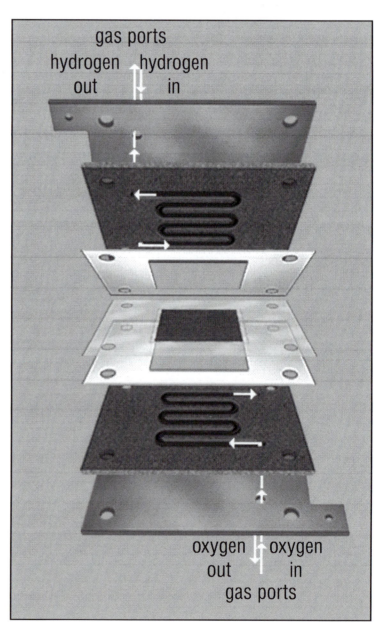

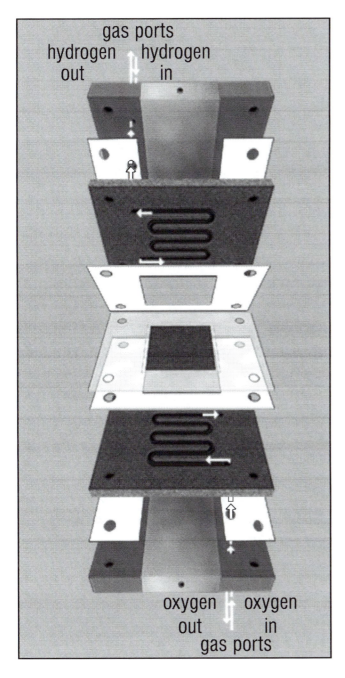

gas ports
hydrogen   hydrogen
out   in

oxygen   oxygen
out   in
gas ports

Layer view of hard graphite single slice oxygen-hydrogen fuel cell. See construction directions for the K18 pages 29-60 .

With individual oxygen-hydrogen cells, an alternative to the serpentine flow field on the oxygen side can be seen on page 118. The oxygen gas grooves slope down, which may help move the water out of the cell. Water traps and special membranes can be used to offset the water accumulation problem, but they make a more complicated fuel cell structure. In a stack configuration, the hydrogen side loses moisture as the gas distributes itself to each cell, and you get the opposite problem – lack of hydration.

Planning for maintenance

If you are designing stacks for practical use, consider maintenance. Can you take the stacks apart and put them together again with ease? In a large stack comprised of many different cells, if one cell goes down and it happens to be in the middle, you will want to be able to get to it easily.

The more complicated a cell becomes, the more can go wrong with it. The more compact a cell becomes, the harder it is to service it. Mobile applications require a small footprint and one needs to be weight conscious, but for stationary applications there are a lot more options. A larger footprint can be acceptable if it will save on maintenance costs and the original cost of construction.

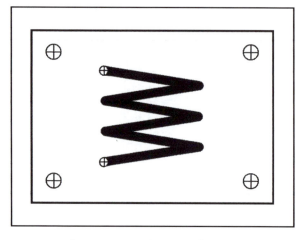

An alternative design for oxygen flow fields or individual oxygen-hydrogen cells

In fact, the fuel cells you will build will probably not require maintenance, but it is always best to think ahead, just in case.

Planar fuel cells

Planar fuel cell configurations address the problem of fitting fuel cells into spaces for smaller electronic equipment such as lap-top computers, etc. In fact, a planar configuration can help to solve the moisture and heat problems of stacks and can help the cells to operate more efficiently. As an aside, a tubular configuration might be even more efficient and maintenance free.

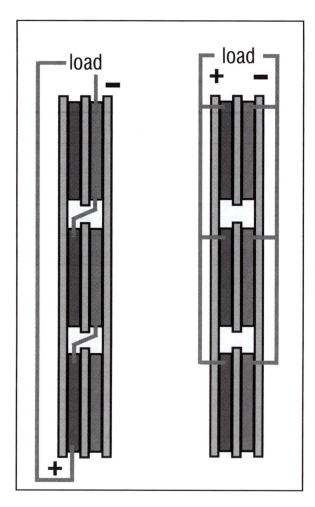

Planar fuel cell connections, left, series; right, parallel

Gas port plate

Another configuration would be to have individual cells plugged into a gas port plate. The gas port plate would have individual feed and outlet channels within it and also a water trap. The mechanism for maintaining hydration on the hydrogen side could be incorporated into such a plate. Such a design would be very practical for maintenance and cost savings.

Stack design considerations – hardware

Even the basic hardware used in fuel cell stacks should be considered carefully. For instance, nylon screws and nuts could be used to bind the stack together to eliminate the need to coat the screws or cover them with shrink tube. For that matter, other methods of tightening or fastening might be employed. In many stack designs the pressure screws are outside the perimeter of the flow field plates so that these plates do not need to be drilled. With such a design, the end plates are the only plates that have holes for the binding screws. However, with this configuration you have to find a good way to align the inner components. With a design such as shown in this book, the fastener holes align and hold fast all of the cell component layers.

Longevity

If you build fuel cells for simple experimental purposes, the longevity or quality of the materials might not make much difference. If you are designing for practical applications and long life, you need to consider the quality and characteristics of the materials used in your design. If you use metal screws, for instance, for longevity you will want to use the best stainless steel screws and nuts available, or nylon, so that you will not have a corrosion problem to deal with later.

Operating environment

Consider the fuel cell's operating environment. If you build a fuel cell stack for your sail boat as a back up power supply, for instance, consider the effects of salt spray. There can also be problems with dissimilar metals that corrode when in contact with each other in the fuel cell due to electrochemical action.

In summary

In summary, there are many ways to configure fuel cells, and many different materials appropriate for constructing them, so it's possible to tailor them for the particular purpose you have in mind. Lightweight and compact is important for mobile applications. Heavy with a larger footprint is acceptable for stationary applications if it will save you time and money for maintenance and initial cost.

Building Stacks

Making a fuel cell stack is very simple. One gas feed line supplies the hydrogen to all the cells in the stack through an offset gas port hole as in the L78. Bipolar plates can be used for the inner plates of a stack to conserve space.

Bipolar plates

A bipolar plate provides the series connection and has gas channels on either side of the plate. One side of the plate has the oxygen feed grooves and the other, the hydrogen feed grooves.

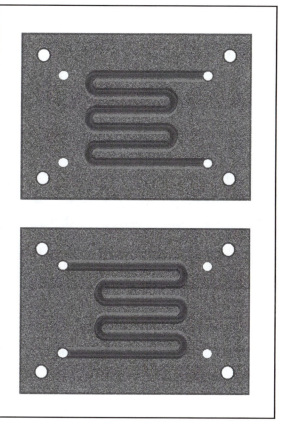

A bipolar plate, front and back.

Plate thickness and groove depth

The graphite plates can be the same thickness as for the single slice cell, in which case the grooves should be more shallow. If the plates are cut a little thicker, the grooves can be the same depth as for the single slice cell. These instructions assume that you will be making your bipolar plates thicker than the single sided plates.

When making the grooves for bipolar plates, be attentive to the groove depth as it is easy to crack the plate. The depth of both the oxygen grooves and the hydrogen grooves can be reduced, or just the hydrogen grooves; or, make the plate a little thicker. If you use the graphite foil for bipolar plates, reduce the depth of the grooves even more than for the hard graphite plates. Experiment to see what works for you.

Gas feed in the gasket

Another way to use bipolar plates is to make the oxygen side gas flow grooves on one side of the graphite, and leave the other side flat. (See template, page 217) For this design, the rubber gasket

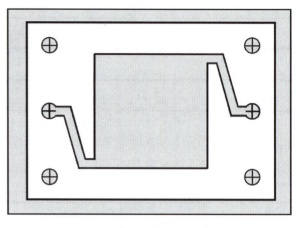

Surround gasket with gas distribution area.

lies against this flat side of the graphite plate and is cut with a channel that feeds the gas into the gas diffusion cloth area (see diagram). With this design there are no serpentine flow fields – just feeder channels from the gas port to the carbon cloth surface. Use either the .020 silicone rubber or any other thickness that allows the graphite plate to connect with the carbon cloth surface correctly.

Stack templates

I have included two sets of templates for stacks, one for a convection stack (pages 202-210), and one for an oxygen-hydrogen stack (pages 211-216). Both are made from hard graphite, and based on the K18 fuel cell design.

Convection stack configuration

The diagram at right shows the layers for a hard graphite convection stack. The outermost graphite plate on each side has gas grooves on one side only. Each additional graphite plate is bipolar (gas grooves on both sides). For each MEA (membrane electrode assembly) added to the first MEA, you would also add one bipolar plate and rubber surround gasket.

Facing page, layer view for a hard graphite convection stack. See construction directions for the K18.

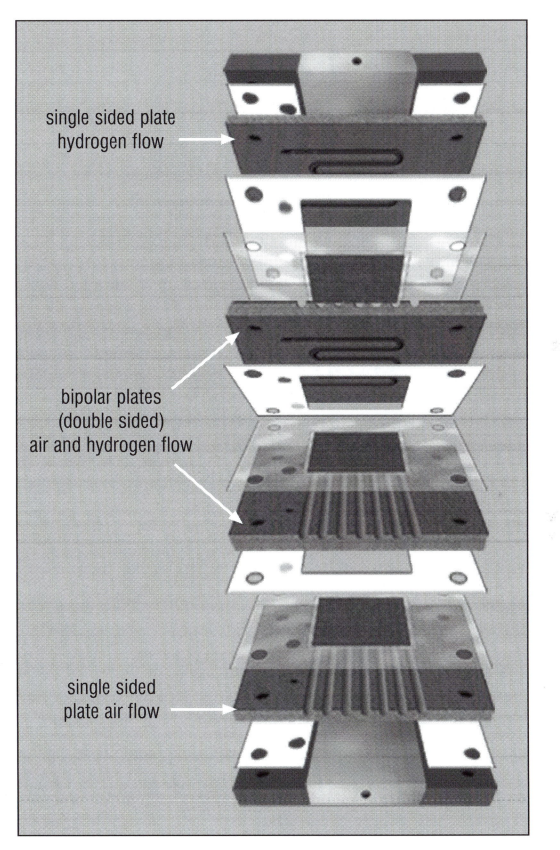

single sided plate
hydrogen flow

bipolar plates
(double sided)
air and hydrogen flow

single sided
plate air flow

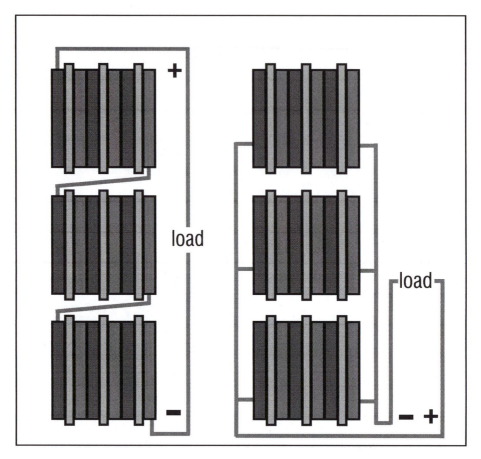

Electrical connections for stacks. Left, series; right, parallel.

Oxygen-hydrogen stack configuration

Oxygen hydrogen stacks can be constructed by following the templates given. There are two gas distribution channels, one for oxygen and one for hydrogen.

The configuration of the components for this oxygen-hydrogen stack, beginning from the hydrogen side, is: hydrogen end plate, an electrode with its two rubber spacers, the single sided hydrogen graphite plate, a rubber surround gasket and then the first MEA in its Mylar surround.

Next is a rubber surround gasket followed by a bipolar graphite plate with the oxygen side facing the MEA. Against the first bipolar plate on the hydrogen side is another rubber gasket, MEA, rubber gasket, bipolar plate, rubber gasket, MEA and so on. You just keep stacking until you reach the desired voltage. After the last MEA and rubber gasket is the oxygen single sided graphite plate, electrode with spacers and oxygen side end plate.

Plan for easy disassembly

For stacks, the components should be easy to assemble, and disassemble (if need be). It may be good to have a slightly looser fit between the screws and the screw holes for the stacks; however, the fit should hold your components in alignment adequately. Instead of screws for long stacks, you can use rod of any material with threaded ends.

Larger stacks need gas pressure

To power larger stacks, you will probably want to build a high output electrolyzer such as described in **Build a Solar Hydrogen Fuel Cell System** and use a low pressure storage system to feed your stacks. Larger stacks may need added pressure to distribute the gases at a proper rate.

Gas flow regulator

For experimenting with gas flow rates, I would suggest an inexpensive surplus flow regulator such as occasionally offered by American Science and Surplus or Herbach and Rademan. Any regulator with an outlet pressure of 0-60 psi per fine tune setting will do. You will basically be working with pressures around 2 psi, and you must be careful to not use too much pressure to avoid the possibility of membrane blowout. This will depend on the size and number of your stacks. Pressure regulators are available from a variety of sources.

Additions to optimize performance

For convective stacks, forced air housings are nice. This is a housing for the fuel cell stack with a low voltage, low current muffin fan to circulate the air so that enough oxygen gets to the cells. For larger stacks, a humidifier may or may not be needed. Heat sinks attached to containers of water, and connected to the fuel cells, can help keep the stacks hydrated. The heat given off from the stack heats the water, which then humidifies the stack. For oxygen-hydrogen stacks, wet proofing can be a critical factor on the oxygen side and wet proofed gas diffusion layers would be appropriate.

Gasket thickness affects performance

Gasket thickness may be varied because, with oxygen hydrogen stacks, there are gaskets on both sides of the MEA, which creates a little more distance between the carbon cloth diffusion layer and the graphite plate surface on the oxygen side. This can result in a contact problem. In convection cells, the MEA lies directly against the oxygen side graphite plate with no gasket between them, so this is not a concern.

The carbon cloth diffusion layer must contact the oxygen side of the next graphite plate. If you do have a contact problem using an .020 gasket, try going to a lighter (thinner) gasket. Some of the MEAs I have purchased had a hydrogen side that stuck out from the surface of the Mylar surround a bit more than the oxygen side did – the oxygen side was more or less flush with the Mylar. Once you have mounted the MEA into its Mylar surround, check to see if one side sticks out more than the other.

McMaster-Carr has thinner gasketing available in silicone. Latex is all right, but not the best for this purpose. Its usable highest temperature is around 182° F, which is on the edge for operating temperature for stacks. Experiment to see what works. An alternative to silicone for gaskets could be something with a little less give, such as very thin fiberglass cloth or another material. The .020 silicone rubber might work anyway with compression applied from the fasteners, but at least be aware that this is something to consider when troubleshooting.

Experimental Methods for Making MEAs

About MEAs

After building a few cells you may want to delve deeper into the mysteries of fuel cell construction, fabrication, and design. The next step is to make your own MEAs (membrane electrode assemblies).

MEA components

To make an MEA, you need a proton exchange membrane material such as Nafion in its solid form; Nafion 5% in liquid form; and a carbon fabric or paper which acts as a conductive carrier substrate for the catalyst, and also acts as a GDL (gas diffusion layer). You also need the catalyst, which in most cases will be platinum. Platinum can be used in several different forms to act as a catalyst. Depending on the technique you use, you may also need lamp black, Bucky balls (Fullerenes), etc., and other materials specific to the project you have in mind.

Applying the Catalyst to the Substrate

1. Sputter diffusion

There are several ways to apply the catalyst layer to the cloth, paper or membrane substrate. The common practice at present is to use the sputter diffusion method to apply the catalyst layer. This is a good technique as very thin films can be deposited on a substrate with a greater or lesser degree of control, making a very precise deposition in terms of thickness and amount of materials used. The method works by bombarding a platinum target with high velocity gaseous argon ions. This dislodges molecules of the platinum which are propelled to the surface to be coated. The sputtering method is not discussed further in this book, but the techniques are certainly within the range of the average experimenter.

Information about sputter deposition of the catalyst can be found in **Procedures in Experimental Physics** by John Strong and the journal **The Bell Jar** (a journal of vacuum technique and related topics for the experimenter). These texts have some practical information about the construction of devices for forming thin films by sputtering.

2. A Benjamin Franklin technique

If sputter diffusion sounds hopelessly technical to you, here is an easier way. I'm a high voltage experimenter and an avid fan of Benjamin Franklin. So what could Benjamin Franklin possibly have to do with fuel cells? Well, nothing, that is, if you don't know what to look for. It occurred to me that some of Franklin's electrical and printing experiments might be relevant in the quest to produce a viable catalyst layer.

In one of his experimental moods, Ben was playing with some gold and silver leaf, a few cut-out cardboard silhouettes and some Leyden jars. What he did was cover the silhouettes with the leaf and place a piece of paper very close to the leaf covered side of each silhouette. He then discharged a series of Leyden jars through the leaf. The high current immediately vaporized the leaf and deposited a perfect representation of the silhouette on the paper in gold or silver.

I have not had a chance to try this yet, but for fuel cell applications, particles of vaporized platinum metal leaf would make the finest coating, perhaps with an extremely large surface catalyst area. You could coat or impregnate a catalyst layer on/in polymer exchange membranes, GDLs (gas diffusion layers), semi-liquid PEM solutions (in the drying phase), or particles such as lamp black or Bucky balls.

Accessible technology

For the average experimenter, this means that with very little equipment one can deposit very large surface area loadings with as little metal as possible, without the expense and outlay for high tech equipment. Not bad – thank you Ben!

The technique is definitely viable. If you wish to pursue it, research Franklin's experiments and pick up a copy of Exploding Wires – Principles, Apparatus, and Experiments by Steve Hansen. Instead of thinking wires when you read Hansen's book, think platinum leaf. Hansen also has a book about constructing an impulse transformer if you want to get a little more sophisticated about applying this technique, but Leyden jars with a small electrostatic generator work just fine. These books are from the Bell Jar Project series.

Resources for the experiment

You could use surplus high voltage generators and capacitors to provide the voltage and current. For basic information about the application of high voltage technique to catalyst layer deposition, the High Voltage Experimenters Handbook is available online.

3. Electrochemical deposition

Another method of catalyst deposition would be electrochemical, such as through electroplating. In electroplating, the object to be plated is connected to the negative terminal of an electrical source, such as a battery or rectifier power supply; and the metal to be deposited is connected to the positive terminal of the power supply, with both immersed in an electrolyte bath of the metal salt that one wishes to plate with. When electrical current flows through the electrolyte bath, atoms of the metal are freed and deposited on the object connected to the negative terminal. The platinum on the positive terminal dissolves in this process and takes the place of the metal in the salts in the electrolyte that have been used to plate the object. This process is within the reach of the average experimenter.

4. Electroless deposition

One can also work with what is called electroless deposition of metals. Information about these electroless techniques are given in the resource section of this book. Electroless deposition allows fine control of deposition of the catalyst layer without an extensive outlay in capital and time. It is a fairly straight forward and simple process and is worth looking into.

5. Mechanical deposition

There are several methods of mechanical deposition that can be employed. As an example, thin metal leaf tends to cling to whatever it is applied to. Take a piece of platinum leaf and lay it atop a carbon fabric or paper and burnish it in with a burnishing tool (that is, a smooth metal stick of some sort) to allow the metal particles to be caught in the carbon fabric fibers or porous layers of the carbon paper. Then coat this with a liquid layer of 5% Nafion solution. When applying the platinum leaf it helps to wet the surface of the carbon fabric and to also dip the burnishing tool in water as you burnish the leaf into the fabric. This allows the metal to break up into extremely fine powder.

Powdering platinum

Another method would be to mix lamp black (fine soot) with the platinum foil (leaf) in a mortar and grind until the foil combines with the lamp black well. Thin leaf has a tendency to break up, so this should be easily accomplished. Finely powdered platinum and or platinum black can be substituted for the platinum foil. To make a more finely divided platinum powder for a carbon platinum mix or just as is for a catalyst layer, the platinum can be ground with

Platinum leaf

salt and honey in a mortar. Then add distilled water to this mixture and let it sit. The powdered platinum will collect on the bottom of the container and the water can be poured off. More distilled water is then added, and repeat the process until you are left with platinum and water only. Then, drain off the excess water and let the platinum powder dry for your use.

Carbon nano-tubes

One can also use Bucky balls (fullerenes or carbon nano-tubes) mixed with the platinum to create more catalyst surface area and thus greater power densities in the finished MEA. This mixture would then be evenly dispersed mechanically over the cloth or paper in the density required, then the cloth and catalyst would be coated with a 5% Nafion solution.

Carbon and platinum mixtures

Mixtures of platinum and carbon black or carbon powder can be purchased ranging from a 1% to 40% platinum load per mix (available from companies such as Alfa Aesar), or you can mix your own. Mechanical methods of deposition have advantages and drawbacks and you will have to experiment to see if this method is appropriate for you.

6. Electrostatic dispersion

If you have experienced static cling or have ever worked with a Van DeGraff generator, you will understand the effect of attraction for an electrostatically charged material to a material with the opposite charge. The material to be coated electrostatically can be set upon a plate with one charge and the materials to be attracted to that substance are connected to the opposite pole of the generator. The lighter powders can thus be attracted to the surface of the conductive carbon cloth or paper and coat it. The coated cloth or paper is then covered with Nafion solution. A liquid formulation of the platinum catalyst can also be applied to the carbon cloth or paper via electrostatic spraying techniques.

7. Simple mixing

Platinum powder, platinum black or various combinations of platinum black and carbon powder could simply be mixed with Nafion solution and painted on the carbon cloth or paper. Concoctions of inks made from these platinum substances could be screen printed or placed on the substrate by brushing.

Platinum inks can also be used in printer cartridges and applied to carbon paper, by some desk-top computer printers.

8. Photochemical

The photochemical deposition technique is a process that any photographer who has ever worked with the platinum or palladium printmaking process is acquainted with. The technique is particularly noteworthy for us in this process because it has a history of development (pun intended). Over the years, this process has been fine tuned for photographic purposes so that the quality control, and reproducibility in average hands is high without a lot of high tech equipment or preparation.

One of the advantages of the platinum printing process is that with a minimal amount of equipment, the loading parameters of the platinum can controlled fairly precisely. This makes it possible to methodically increase or decrease the loading to see how low or high you can go with power density given the output requirements you have.

Platinum black

The result of this photochemical process is a fine coating of platinum black. Platinum black is one of the best forms for a platinum catalyst for fuel cell purposes. It is easy to control the strength of the platinum solution by the photochemical method and thus to control the loading; and the solution is easy to apply since it can be hand brushed onto the carbon cloth or paper. This is an excellent low tech application method at minimal cost.

Application techniques

The solutions can also be applied to the substrate in other ways, such as screen printing for more uniform application. Screen printing techniques have allowed the solar cell industry to produce low cost, high power density solar cells and I presume that this technique may presently be used in the fuel cell industry to some degree.

Brewster's reaction

Photochemical application relies on an iron based process and what is termed Brewster's Reaction. In this process, potassium tetrachloroplatinate (a platinum salt) is combined with an iron based sensitizer, either ferric oxalate or ammonium ferric oxalate and exposed to ultraviolet light. Upon exposure to the UV spectrum, the ferric oxalate changes into ferrous oxalate which has the unique characteristic of reducing the noble metals such as platinum from their double salts and then reverting back to ferric oxalate. The ferric salts are then washed from the substrate and you are left at the end of the process with extremely finely divided platinum particles that act as the most superb catalytic agent – platinum black.

Methods to get platinum black

There are several photochemical methods to produce platinum black.

The DOP process, which simply means develop out process, uses a developer. The POP process, (printing out process) uses no developer and the platinum black prints out fully with exposure. We will discuss the POP process as it is easy, requires fewer chemicals and with some modifications is suitable for adhering catalyst layers to carbon cloth substrates. I have experimented with this technique and think it has good potential for producing high quality coatings for fuel cell purposes.

MEA Tool and Materials Lists

MEA tools

Candy thermometer.

Art brushes (small).

Plastic or glass work surface. For coating carbon cloth

Plastic clips. To hang dry carbon cloth

One-burner electrical heating unit with temperature control. Can be purchased at discount retail stores.

Stirring rod. Polyethylene or glass

Pint canning jars.

Small water bath set-up.

Oven and oven thermometer.

Aluminum metal plates. To press MEAs with, about 5³/4" x 5³/4"

Graphite powder. If you constructed the K18 you will have plenty of graphite powder, or you can use ground up pencil lead or purchase powdered graphite from any chemical supply house or Skylighter

4 C-clamps.

Polyethylene lid. From two pound coffee can or other device to help hold down membrane while in liquid baths

Polyethylene tubing. (optional) To help hold down membrane while in baths. Can be purchased at hardware store

Safety gear. Safety gear for all processes and chemicals are described in MSDS (Material Safety Data Sheets). Do not work with acids without rubber gloves and proper protective clothing. Wear a face shield or safety glasses when working with acids. Respirators are also recommended when working with certain chemicals. Read all MSDSs for all chemicals to be worked with. Face shields can be purchased at hardware stores for about $14. Safety glasses are also available at hardware stores as well as respirators and face masks etc. You can also find safety gear at Plastics Industrial & Laboratory Supplies, Lab Safety Supply/Grainger or Gempler's.

White gloves, cotton or nylon. Available at a variety of online sources.

Nick sander (optional). For experimentally roughening the surface on either Nafion membrane or carbon cloth. Available at a variety of online sources.

Bulb-type electrolyte siphon. Available at any hardware, discount, or auto store (used for adding and removing electrolyte from batteries).

The following tools are needed only if you are going to try the experimental method for loading the catalyst:

2 pipettes. 1 ml with 0.01 divisions. Can be purchased from a variety of online sources. At minimum you need two pipettes and I suggest having three to five pipettes so that if you break one or are performing an operation where you will need several, you will have them on hand without having to wash one out immediately.

Pipette pump. Can be purchased from a variety of sources online.

Small plastic container. Rubber Maid, etc., about 12 oz., more or less

UV source. The sun is the best UV source. You can also use a bank of inexpensive UV fluorescent bulbs.

Shot glasses, or other small glass containers.

Fan.

Safe light. Either a yellow bug light or sodium vapor light. Yellow bug lights are less expensive than sodium vapor bulbs.

Paper candy cups. For weighing

Light-tight container. A cookie tin, for instance.

Needle and thread.

Triple beam balance scale or equivalent with 0.01g sensitivity. Edmund Scientific Company. This item can also be purchased through other suppliers. Check the internet for best prices.

MEA materials list

Hydrogen peroxide 3%. Available at any drug store

Sulfuric acid. Can be purchased at most hardware stores in the plumbing section, or auto parts store as battery electrolyte, or any chemical supply house.

Distilled water.

Nafion NRE-212 membrane. Can be purchased from Fuel Cells Etc, Fuel Cell Earth, Alfa Aesar, Sigma Aldrich, Ion Power and other online sources. Nafion NRE-212 is 0.002 in. thick. You can use other thicknesses such as Nafion 115 (0.005 in. thick), or Nafion 117 (0.007 in. thick). Also can be purchased at other suppliers.

Nafion solution. Can be purchased from Alfa Aesar, stock #42117, 25 ml Perfluorosulfonic acid-PTFE copolymer, 5% w/w solution. Can also be purchased from other suppliers.

Silicone rubber adhesive. Available at any hardware store. All Electronics offers an electronics grade.

Catalyst. Catalyst in powder form or loaded onto carbon cloth or paper is available from Alfa Aesar, Sigma Aldrich, Fuel Cells Etc, Fuel Cell Earth, Ion Power and other online sources.

The following materials are needed only if you are going to try the experimental method for loading the catalyst:

Ammonium Ferric Oxalate (AFO). $(NH_4)_3Fe(C_2O_4)_3 \cdot XH_2O$ can be purchased from Bostick & Sullivan, stock #AFOD, or other chemical supply house such as Alfa Aesar. It can also be purchased from Photographers Formulary, stock #10-0502.

Potassium Tetrachloroplatinate (II). Can be purchased from Alfa Aesar, stock #11048. Also available from Bostick & Sullivan and other platinum process suppliers. Please note that you want to use tetrachloroplatinate (II) K_2PTCl_4. You do not want to use K_2PTCl_6 which is potassium hexachloroplatinate (IV).

Carbon Cloth or Paper. Can be purchased at Fuel Cells Etc, Fuel Cell Earth, Alfa Aesar, Ion Power, Sigma Aldrich. Wet-proofed cloth is also available, but the wet-proofing may interfere with the catalyst loading method suggested here. You can also purchase carbon cloth at other suppliers listed for carbon cloth in the back of the book.

Loading the Catalyst - an Experimental Method

Making percent solutions

For the platinum process you will have to make percent solutions with chemicals and distilled water. This is really very easy. For instance a 10% solution consists of 10 grams of the solid substance mixed with enough distilled water to make 100 milliliters of the solution. Note that I did not say to take 10 grams and then add 100 milliliters of distilled water. You need to add the distilled water to make 100 milliliters.

In the formulas listed below for the reduced amount of platinum process chemicals used for the necessary first demonstration, I cheated a bit, so it is not this accurate, but it works for our purposes. However, for absolute accuracy, follow correct procedure and keep notes on exact measurements. Search the web for more information about how to make percentage solutions by weight and volume. This will come in handy as you will be working with very small quantities. For instance, for 13 milliliters of a 35% solution, you must figure out just how much chemical substance and distilled water you will need.

Practice the techniques

Practice all the procedures, for instance, using the pipettes with the pump, and make sure you have all equipment and materials properly situated before attempting to actually do the process. You can avoid a lot of surprises and mistakes by doing this. Keep a clean work area and properly dispose of materials. Platinum photo process chemicals are toxic, so be cautious in handling any of these substances. Sulfuric acid is dangerous, so definitely read the MSDS and absolutely do not work with this acid without eye protection and protective hand gear and clothing.

MEA components

The average MEA (membrane electrode assembly) consists of a proton exchange membrane such as Nafion sandwiched between two pieces of carbon cloth or carbon paper. The carbon cloth or paper is coated on one side (the side that will face the membrane) with a catalyst such as platinum, and is conductive. It is called a GDL (gas diffusion layer). The GDL assists in the diffusion of the gas to the platinum catalyst and assists in water movement to and from the catalyst area.

The substrate and catalyst

To increase hydrophobicity (wet-proofing), the carbon cloth or paper can be coated with Teflon. In stacks this can be particularly important, since water buildup on the oxygen side can be a problem. (However, wet proofing may interfere with the catalyst loading method suggested in this book.)

The platinum catalyst layer traditionally consists of 10 wt% Pt particles that are from 2.5 to 8 mm in diameter on a carbon support such as Vulcan XC-72. The platinum catalyst can be applied to the substrate in a variety of ways, such as screen printing, painting on with a brush or airbrush, etc.

The membrane

The proton exchange membrane can consist of any number of commercially available membrane materials such as Nafion. Although Nafion 117 was widely used, the trend is now toward thinner membranes such as Nafion NRE-212 as they provide greater power density than the thicker membranes. There are also other materials and methods of construction for MEAs. For instance the proton exchange membrane itself can be coated with the platinum catalyst instead of or in addition to the GDL.

Construction overview

The steps for making an MEA are:

1. Prepare the conductive gas diffusion catalyst carrier.

2. Clean and condition the proton exchange membrane.

3. Hot press the membrane and catalyst carrier together.

4. Set the components in a Mylar or similar type holding frame.

Planning to load the catalyst

Before you embark on the whole process of loading the catalyst, you will need to plan the timing of this venture. The AFO solution must be prepared 24 hours before it is mixed with the platinum solution; and then, once the coating solution has been applied and the cloth is completely dried, it needs to be exposed to strong sunlight – full sun at midday in the summer is best. If you have UV light tubes and a UV exposure box, you do not need to be concerned about this; otherwise, you will want to monitor weather reports to try to time the process so that you'll have good sun to expose the coated GDL.

Carbon cloth substrate

The substrate for a gas diffusion catalyst holder must be a conductive material such as carbon cloth or porous carbon paper. I have not yet experimented with carbon paper, and would suggest with these techniques to use carbon cloth first and then try carbon paper after you have made a few MEAs and reproduced these results.

I used a 1¾ ounce plain weave carbon cloth. There are products avail-

1¾" plain weave carbon cloth.

Left, Toray carbon paper; right, carbon cloth.

able such as Toray carbon paper and carbon cloth that are either wet proofed or plain. I used the plain, without the wet proofing. Wet proofing simply consists of spraying or coating the fabric with Teflon. This is an advantage on the oxygen side of the cell where water vapor forms . Use the non-coated for your first experiments so that you can reproduce my results, and then move on and experiment with different types of substrates with different coatings if you wish. Store the cloth in its package to protect it from contaminants.

Cutting the cloth

To avoid getting oils from your hands and other contaminants on the cloth, when you work with it, wear thin white cotton or nylon gloves. Cut two pieces of carbon cloth. Each piece should be 1³/₈" by 1³/₈", and should be cut square on the grain. Later, the Nafion membrane will be sandwiched between these two pieces of carbon cloth.

The opposite sides of the fabric look identical, and later in the process it will be important to have some way to tell which side the catalyst was applied to. An easy way to mark the fabric is with a needle and colored thread. Tie a knot in the colored thread on one side of the fabric and remember that the knot side is the catalyst side.

When the carbon cloth pieces are cut and marked, put them in a small plastic bag or other receptacle to protect them from dust, etc. The carbon cloth has a tendency to unravel at the edges. This is not a major problem, but handle the edges with care.

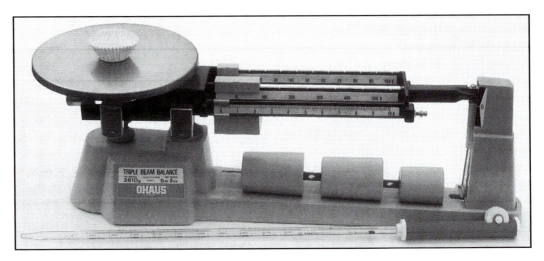

Scale with glassine "candy cup" for weighing chemicals.
Pipette with pump is in the foreground.

Weighing the chemicals

Next, prepare a 35% solution of AFO (ammonium ferric oxalate). You should have a scale with 0.01g sensitivity to measure small quantities of chemicals for this process. I already had a scale with only a 0.1 sensitivity so I had to interpolate. If you have something like this you can use it, but if you are going to purchase a scale, go for the 0.01g sensitivity.

Plastic spoons are good to scoop the chemicals out of the containers. Do not use metal spoons for the chemicals used in this process, and use a separate spoon for each chemical. You will also need measuring containers to put on the scale to hold the chemicals you weigh. I use 1" "candy and party cups" made of a glassine paper. The paper is not porous so pouring the weighed powders out is very easy since nothing sticks to the sides. You could also use a small glass, or the small plastic cups used for sauce. Just remember to compensate for the weight of the container (tare) when you zero the scale. If the scale is new to you, practice using it before you start making the chemical preparations for this process.

Safety precautions

Be sure to follow all MSDS and safety precautions for the chemicals you are using. Safety glasses should be worn for any chemical work. Always wear a respirator and use protective gloves when handling chemicals. If you use disposable latex gloves, put them on and then run your hands under water to remove the powder that comes on the gloves so that it does not contaminate your work. They can be purchased without the powder coating but the most common kind has it.

It is your responsibility to read all the MSDSs for the chemicals you use and obtain proper protection and follow safe storage and use procedures as well as safe disposal procedures. If you know any photographers who work with the platinum printing process, talk with them about how they work safely with the materials involved, and to learn a little about the process.

The work area

Your work surface should be easy to clean if you spill the contents of any materials, and you should have clean up materials handy. You could build a small fume hood which is simply a five sided box with the open end facing you and at the top or back of the box, a fan with flexible duct to suck out any particles that may become airborne. The inside of the box is your work area and should be large enough so to perform whatever operations you need to within its confines.

Also make sure children and pets will not be able to get at any of the items you use; and properly dispose of all contaminated materials you use (such as plastic spoons, etc.) immediately after you complete your tasks.

AFO preparation

The equipment and environment needed to prepare AFO solution include a light tight room, a safe light, scale, paper cups for weighing, plastic spoons. You will also need distilled water, pipettes, pipette pump, one shot glass with Saran Wrap and rubber band, a plastic or glass mixing rod, light tight container for storage.

Safelight conditions

The AFO (ammonium ferric oxalate) must be prepared in safelight conditions – either a sodium vapor light or yellow bug light in a light-tight room. If you happen to have access to a dark room, that's perfect. To prepare this solution, I wait until after dark, and then use a yellow bug light for illumination to weigh out the AFO and make the solution.

Timing and storage

Prepare the AFO solution at least 24 hours before using it, and after preparation, the solution must be stored in a light tight container of some sort. I used a cookie tin. Remember, the AFO can only be removed from the light tight container and worked with in safe light conditions (yellow bug light or sodium vapor light). When storing the solution it should be kept cool, but it does not have to be refrigerated. Light, heat and time may convert ferric oxalate to ferrous form. Storage time can be up to several months.

Making the solution

Prepare a 35% solution of AFO. To do this as accurately as possible, use a pipette with .01 ml gradations and a pipette pump (a small plastic device that fits on the end of the pipette). As with the scale, practice using a pipette with the pump before you actually make your solution. When you are working with a pipette it is sometimes easier to see the gradations if you hold a piece of white paper behind the pipette as you draw up the distilled water.

Draw out .55 ml of distilled water with the pipette and drop into a shot glass or other small glass or plastic container. Do this in a well lit room as it is hard to see the tiny gradation scale on the pipette in the yellow safe light. Pre-adjust your scale in regular light also, before you weigh out your material in the safe light.

After you have put the distilled water in the shot glass and adjusted the scale, turn off all lights and turn on the safe light in your light-secure room. With the safe light on, weigh out .3 grams of AFO and then pour the powder into the shot glass with the distilled water. Swish it around a bit and it should mix well, or use a stirring rod (the AFO mixes easily with water). When the solution is mixed, cover the shot glass with a small piece of Saran Wrap and put a rubber band around it to hold it tight. Put it into your cookie tin or other light tight container.

Handling AFO

Be sure to replace the lid tightly on the AFO bottle that you purchased. Never open an AFO bottle in other than safe light conditions. Also, do not open the AFO bottle to see what it looks like when you first buy it unless you are in safe light conditions. After you have all your AFO light-secure, you can turn off the yellow light, and store your light-tight box (cookie tin) in a cool place, out of reach of children. The 35% solution of sensitizer will be ready to use in 24 hours.

The amounts of sensitizer and metal salt solution that you will be making are very small as it does not take very much to coat the two pieces of fabric. This weight of mix will give you .85 ml of AFO solution.

Procedure summary

To reiterate the procedure – draw out .55 ml distilled water into shot glass and adjust the scale in normal light. Make your work room light secure and turn on the safe light. Weigh out .3 grams of AFO into shot glass. Mix or swish solution, cover with plastic wrap and secure with rubber band. Put in a light tight container, store in a cool area and wait 24 hours before use. This procedure takes about two minutes to do and is very easy.

Preparing the platinum solution

Do not prepare the platinum solution until you are ready to mix it with the AFO for immediate coating of the substrate. The platinum solution does not have to be prepared in safe light. You will need the scale and paper cups for weighing, a shot glass, stirring rod, hot plate with adjustable temperature control, water bath, warm distilled water, and a candy thermometer. Any heat source that you can regulate can be used.

Water bath

A water bath is very simply a pan of water with a metal trivet placed in it to set the shot glass on to be heated. The trivet elevates the shot glass so that it does not come into contact with the direct heat from the burner on bottom of the pan. Do not apply direct heat to warm your solutions, always use a water bath. The water level in the bath should be part-way up the sides of the shot glass, but not enough for the shot glass to float and become unstable.

To prepare a 33.9% solution of metal salts, heat the water bath and attach the candy thermometer to the side of the pan so that you can easily read the temperature as it heats up. Let the temperature rise to around 110° or 120°F. Do not heat to 140°F as the chloroplatinate will turn to another form of chloroplatinate which you do not want. Warm a shot glass full of distilled water in the water bath.

Making the platinum solution

When the water is warmed up to temperature, take the shot glass of distilled water out of the water bath and set it aside. Immediately weigh out .2 grams of tetrachloroplatinate, and put it into another shot glass and place it in the water bath. Next draw out .39 ml of warm water with the pipette from the shot glass of distilled water that you just took out of the bath. It should still be warm. Drop the water into the shot glass containing the tetrachloroplatinate. Swish the shot glass of tetrachloroplatinate solution around to help dissolve the powder into solution. Keep swishing it until all the tetrachloroplatinate dissolves. If you take it out of the bath to mix it, do not keep it out for too long or it will cool down. If the platinum solution does not appear to be dissolving after a short while turn up the heat, but don't let it go to 140°F.

What you are doing is mixing a platinum compound with water to form a solution. Although the process is not instant, you do want to get the platinum in solution as fast as possible – but make sure it is fully dissolved. Another point is that heated water reacts differently in a pipette than water at room temperature. The heated water tends to spurt out like a fountain by itself before you are ready to release it, so practice drawing up and then releasing warm water ahead of time. The tetrachloroplatinate should dissolve within one or two minutes. The warmer the water that you pipette into the shot glass, the faster it will dissolve the chloroplatinate. The whole operation takes about five minutes to do.

With the tetrachloroplatinate in solution, turn down the heat to about 90°-100° F to keep the solution warm. Then, in safe-light conditions, mix the coating solution within ten minutes of making the tetrachloroplatinate solution. To do this, take the AFO from the light-tight container and pour it into the shot glass containing the tetrachloroplatinate solution. Swish it to mix or stir it a little with the end of a plastic handled paint brush.

Use a brush with no metal

Once mixed, the two solutions are ready to apply to the carbon cloth. For this, use a cheap (yes I said cheap) art plastic paint brush with no metal on it. It seems that all the expensive brushes have a metal rim to help hold the fibers in place. Do not use these as the metal will react with the platinum salts.

If you cannot find all-plastic handled brushes, take some crazy glue, go around the rim where the metal comes into contact with the bristles, squirting the glue into the bristles around the rim. When it dries, the fibers will bond together and prevent the solution from wicking up to the metal. Some platinum printers use a Hakke brush which you can get at art supply stores, but the ones I have seen are too large for this application.

Coating the carbon cloth

Before coating, put the brush in a container of distilled water to wet it and leave it there until you are ready to coat. I usually coat on a thin sheet of Plexiglas but any smooth surface such as glass will do. When you are coating, some of the coating solution goes straight through. I move the piece and, with the brush, dab up the solution that goes through to the Plexiglas surface and add it back to the surface of the carbon cloth. A smooth Plexiglas or glass surface will not soak up the solution and thus not waste it.

When you are ready to coat, remove the brush from the water and squeeze the water from the bristles with a paper towel. The reason for keeping the brush wet and then squeezing the water is to wet the fibers so that they do not soak up and waste coating solution.

Dip the brush into the coating solution and apply the coating to the cloth (on the side marked with a knot) with quick rapid strokes up and down, and then across, to assure thorough coverage. Try to apply equal amounts of the solution to each piece of cloth.

Make a test paper

Coat a small piece of paper with some of this solution as a test piece. When the test paper is exposed to the UV source, it will turn black and tell you that your chemistry was correct. You will also be able to watch the development as it proceeds while exposed to UV rays. The black is very finely divided particles of platinum, and when you see it you will know that you have found the mysterious lady – Platinum Black.

Dry the fabric pieces

After coating the fabric, hang the pieces up to air dry. As you lift the pieces off the Plexiglas surface some of the solution will drip down to the bottom edge of the piece. You can let it drip for a few seconds and then put the fabric back on the Plexiglas and coat it some more with the solution that dripped off. Use plastic clips (do not use metal) to hang the cloth and use a fan to blow gently on the hanging pieces.

In the POP process, the fabric pieces are dried to ambient humidity, so don't use a hair dryer or any applied heat to dry them – use moving air at ambient temperature. Let the pieces dry in the dark in your light-tight area. If you have to work in the area where they are drying, use safe light only. If you leave them to dry, turn off the safe light. Let them dry for at least several hours. The coated fabric should not be exposed to UV until it is totally dry.

Expose to UV

After the pieces have dried, they can be exposed to the sun outdoors. The sun is a low cost and very effective UV source. The best conditions for exposure are during summer between 10am and 2pm without any clouds in the sky. Check the weather forecast ahead of time. The cloudier or hazier the day, the longer the exposure time will be.

If you do not want to use sunlight, you can purchase UV fluorescent light tubes and build a UV exposure box.

Take the fabric outside for exposure and face it (the side marked with the knotted thread) directly

Exposing the catalyst to UV

toward the sun. I usually clip the fabric to a Plexiglas panel on a frame that can be angled to take best advantage of the position of the sun. Expose the fabric for as long as you can. Unlike the photographic process, for platinum catalyst loading over-exposure is not a concern. The fabric should get as much exposure as possible so that the ferrous oxalate will reduce the salts to as much platinum as possible. Four to six hours is good, however, exposure times of an hour are sufficient. On your first try, give it as much exposure time as possible.

Washing out the ferrous salts

After exposure, fill a small plastic or glass receptacle with distilled water and place the cloth gently in the water. The orange-yellow powder that had formed on the surface during drying and exposure is washed out into the water. You may notice some platinum washing out also. After about two minutes, drain out the water and fill the receptacle again with more distilled water. Let it sit for another two minutes. This should sufficiently clear the piece. Wash each piece separately.

When you pour the distilled water into the wash receptacle do not pour the water over the cloth as you do not want to disturb or dislodge platinum particles from the fabric. Pour the water into the side of the dish very gently. Make sure the fabric is totally immersed and not floating. Two washes seem to be sufficient. After washing, hang the pieces up to dry for several hours and then store them in a plastic bag or non-metallic container to protect them.

Other washing methods

I have deviated from the traditional method of washing. You can use more water clearing baths. Platinum prints are usually washed in flowing water for two minutes and then subjected to water baths as I had described; then washed again in gentle flowing water to clear the prints of any ferrous salts.

Other agents can be used for clearing such as a weak solution of hydrochloric acid, or phosphoric acid, or EDTA can be used along with the previously mentioned chemicals.

Once you have loaded the GDL (gas diffusion layer) with the catalyst, you are ready to prepare the proton exchange membrane.

Preparing the Proton Exchange Membrane (PEM)

Membrane thickness

There is a range of membrane thicknesses that can be used. I used a Nafion NRE-212 membrane purchased from Alfa Aesar. I figured that I would get greater power density from using a thinner NRE-212 membrane, and it was less expensive than purchasing the thicker 115, or 117. I do think the thicker membranes would be more suitable for some circumstances.

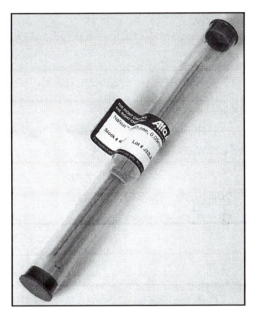

Nafion membrane rolled up in a tube

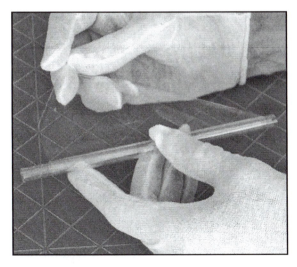

Handle the Nafion membrane very carefully

Handling the membrane

When you handle the membrane, use white gloves so you do not contaminate it with your skin oils, etc. If you order your membrane from Alfa Aesar, it comes to you in a plastic tube with the membrane inside the tube wrapped around another plastic tube.

Membrane preparation

On the PEM material, mark a square 1⅞" by 1⅞" with a ruler and fine felt-tip pen. Carefully cut it with scissors, rotary cutter, or whatever.

Bathing the membrane

For this process you will need several pint canning jars. Try to get the jars with wide mouths, and milliliters and ounces marked on the outside. You will need your hotplate and water bath or double boiler setup, and a candy thermometer, a 3% bottle of hydrogen peroxide and a bottle of sulfuric acid. The membrane will go into in six successive baths. The first bath will be distilled water; the second, hydrogen peroxide; the third bath, sulfuric acid. The last three baths will be distilled water. It is best to do the hydrogen peroxide and sulfuric acid baths outdoors if you do not have a fume hood. I do much of my work outdoors in good weather.

Keep a steady temperature

The membrane is immersed in each of these baths for one hour at a temperature of 176°F. Clip the candy thermometer to the side of the pot and keep the temperature steady. Do not leave any part of this process unattended – especially when working with the sulfuric acid.

Safety precautions

Be sure to wear proper protective clothing, rubber acid-proof gloves and a face shield. Sulfuric acid does nasty things to organic matter so be very careful and definitely wear protective gear when working with this acid. Even if you think you are the most careful person in the world, don't do something stupid like not wearing protective gear. For rubber work gear, see the resources section of this book. You can also use any rubber type clothing you have such as a raincoat, etc. Again – do not handle this acid without proper safety gear. Read the MSDS before working with this acid and or any other chemical listed in this book.

Keeping the membrane immersed

When you immerse the membrane in the liquid baths it will tend to curl and float. Keep it totally immersed while in the baths. To keep it from floating, I cut a polyethylene holder from a coffee can top to have

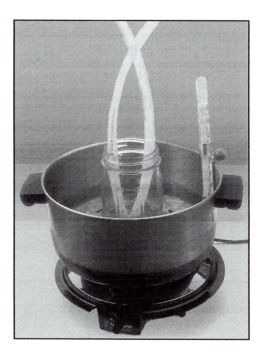

Water bath

feet to raise it from the bottom when placed in the jar, and I poked holes in it to release steam and water. Even this device tended to float so I had to put a piece of polyethylene tubing to hold that in place, as shown in the photo. Use any device that works, but it must be a non-reactive substance such as polyethylene or glass.

Put the piece of Nafion in the water and put the device on top of it to keep it immersed. Be sure that the Nafion is covered with liquid. With a little effort you can probably come up with something better to keep the membrane under water.

The first two baths

The first bath will be distilled water. Fill a canning jar with 100 to 200 milliliters of distilled water and drop the membrane in. Insert the device to hold it down. Let it hydrate in the bath at 176°F for one hour. Before that hour is up, pour 100-200 ml hydrogen peroxide into another canning jar and put it in the water bath next to the one with the distilled water so that it can get up to heat. Transfer the membrane and holding device from the first jar into the second and keep it in for one hour at 176°F.

Fishing for the invisible membrane

When the membrane is submerged you will not be able to see it in the water at certain angles – it becomes practically invisible. When you try to fish it out you will be sort of fishing blind. I use a wooden shish kebab stick to fish for the membrane very carefully. If you poke a hole in the membrane it will be useless, so fish it out with the blunt end of the stick and lift it gently out of the water to transfer it to the hydrogen peroxide. It is also very slippery, so be sure that the jars are up against each other when you do this as the membrane may fall into the water bath or on the ground. After a few times you will get the knack of fishing it out and transferring it. If you can find another device to do this job better, then use it, but do not use anything made of metal.

The third bath

Just before the hydrogen peroxide bath is done, take the canning jar you first used, dry it if it is wet and put the sulfuric acid into it with a rubber bulb siphon tube. Siphon up the acid from the bottle and transfer it carefully into the canning jar. Put in 100 to 200 ml, and put this jar next to the jar in the water bath.

When the peroxide bath time is up, lift the PEM from the jar and let as much of the peroxide drain from it as you can, and then place it into the jar with the sulfuric acid. If you are using a device similar to the one I made to hold the pembrane in place, position it over the pembrane and use the tubing to push it down. Do not use the wooden BBQ stick to do this. You could also use a poly stick or glass rod as the acid will not react with these materials.

Last three baths

Let this bath go for an hour at 176°F. When this is done, take the canning jar out of the water bath, remove the membrane from the jar and let it drain a bit. Then put it in the next jar of distilled water in the bath. Remove the acid with the bulb siphon tube, put it in a jar or poly bottle, put the lid or cap on it securely, and put it aside in a safe place until you can dispose of it properly.

After an hour in this water bath, there are two more baths of distilled water to do. When the six baths are completed, take the membrane out of the last bath and place it on a flat surface to dry. I use a piece of plastic screen (do not use metal screen) in a frame so that it can get air and dry on both sides quickly. Uncurl it and lay it out as flat as possible being careful not to puncture the membrane. When this is dry it is ready. The next step is to hot press the components of the MEA.

Hot Pressing the MEA

Hot pressing the MEA is easy. You will need liquid nafion 5% solution, an oven (do not use a microwave oven), a few stiff metal plates, an oven thermometer, four c-clamps, some graphite powder or pencil lead, and some gloves. An hour before you are going to hot press, coat the catalyst side of the carbon fabric with a 5% nafion solution and let it dry. Don't overcoat or undercoat, and try to coat the piece evenly. The nafion solution does not need to drip through the fabric; it should just coat the surface.

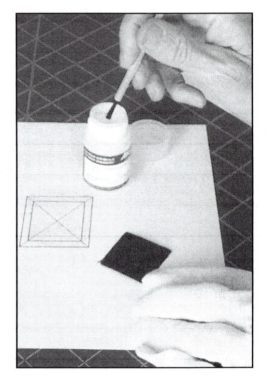

Coat the loaded carbon cloth with Nafion solution

Pressing plates

For pressing plates I used a few small pieces of bronze or brass sheet that was thick and stiff, about 1/16" thick and 5 3/4" by 5 3/4". Since I was not sure what the composition of this metal was I went to the local hardware store and picked up a small sheet of thick aluminum flashing and cut two pieces from that the same size to serve as inner plates to press against the membrane sandwich. However, it would be better to work with one pair of plates – I just wanted to use the sheet metal I had on hand.

Aluminum is the best material for this and should be at least 1/16" thick. If you have a metal shop nearby you should be able to get a few pieces cut for you to 5 3/4" square, or it can even be smaller. Look around to see what you can find.

*Lay out the MEA layers in the middle
of one plate.*

Coat the plates with powdered graphite

I coated one side of each aluminum plate with powdered graphite left over from machining graphite gas flow plates. Instead, you can purchase powdered graphite from the suppliers listed in the back of this book, or use soft pencil lead and scribble the graphite on the aluminum.

Rub the graphite powder into the surface of the aluminum plate and then lightly rub it off with a rag. This leaves some of the graphite on but removes the loose powder. The rubbed graphite side of the plate will be up against the MEA. Lay the aluminum plate down with the rubbed graphite surface upward. The graphite keeps the MEA from sticking to the aluminum plate during the heating and pressing process.

Lay out the MEA layers

Next lay one of the carbon cloth pieces platinum side up on the aluminum plate, and position the membrane over it, making sure that it overlaps evenly on each of the four sides. Then, lay the other piece of carbon cloth with the platinized surface face down on top of the membrane, lined up squarely on top of the other carbon cloth. Next, carefully lay the other aluminum plate, rubbed graphite side down facing the MEA.

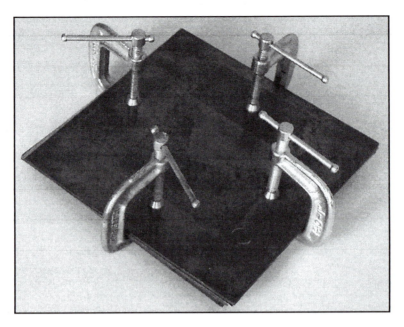

The MEA is clamped between the plates

Add the top plate carefully

It is very important to be able to place that top plate without moving the cloth and PEM combination after they are lined up. The surface is slippery and they will move out of alignment with each other very easily. If this happens, you won't know it until you're finished pressing them, and you will have a useless MEA. So, be diligent when you put these parts together. It would be a good idea to practice this process before doing it so that you can develop the best way to position your plates without moving the membrane sandwich.

Clamp the sandwich

Carefully tighten one c-clamp on the sandwich lightly, just enough to keep the pembrane cloth combination from moving. If you have the plates partially hanging over the table edge, you can put on the first c-clamp and tighten it without moving the plates. Add the other 3 clamps and tighten firmly, but not enough to really squish the sandwich.

Cook the sandwich

When all four clamps are secured, you are ready to heat the sandwich. Preheat your oven to 194°F – use an oven thermometer. Put the clamped plates into the 194°F oven and leave them there for an hour. Then, raise the temperature to 266°F gradually over the next 30 minutes. This is called the glass transition temperature (T_g) for the Nafion membrane. Amorphous polymers such as Nafion have exact temperatures at which the polymer chains go through a big change in mobility. This added mobility plus pressure creates a good sandwich. (Please note that the glass transition temperature has nothing to do with melting temperature and the two should not be confused.)

Increase the pressure

When the temperature reaches 266°F, quickly take the sandwich out of the oven with protective gloves on, and tighten all four clamps by hand, tightening one a little, then the next clamp a little, continuing around in a circle until they are all as tight as possible. Put the sandwich back into the oven and note what the temperature is. More than likely the temperature dropped and you will have to get it back up to 266°F as fast as possible. Turn up the oven to 300° or 400° F to try to bring the temperature back up to 266° as fast as possible.

When the temperature reaches 266°, turn your oven control down to about 250° or whatever it takes to maintain the 266°F for two minutes. After two minutes at 266°F take the sandwich out of the oven, loosen the clamps slightly to relieve pressure on the MEA, and let it cool to room temperature. This will take about one half hour. Undo the clamps, remove the plates and you will find that the cloth has bonded with the

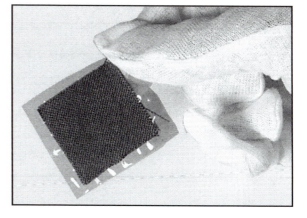

The completed MEA sandwich

membrane. The MEA (membrane electrode assembly) is ready for insertion into the Mylar surround.

Once you get the basics down you can begin to experiment with different platinum loadings and work with other variables. To summarize the above technique: a 33.9% solution of potassium tetrachloroplatinate, and a 35% solution of ammonium ferric oxalate are combined to make the coating solution. These two solutions are combined in a one to one ratio – for every drop of platinum solution you add one drop of AFO solution.

Further Experiments

Lower the loading

If you want to experiment with different levels of loading you only need to add more distilled water to the final coating solution which will lower the amount of platinum spread over a certain fixed surface area. In this manner you can test to see how reducing the amount of platinum over a fixed area affects the power output of your fuel cells.

Increase the loading

To increase the loading over a specific area, just brush on more coating solution. One way to do this is to coat the surface with, for example, double the amount given in the above formula. Another method would be to give it one coat with the amount given above, let it dry to a damp stage but not completely dry, and then put another full coat on. One can continue this process for as many coats as desired.

Another way would be to do one regular coat, let it dry, expose it and wash; and then apply another coat, expose and wash, and so on. You can also work with heavier carbon fabrics; or, try coating the Nafion membrane also. I have not tried this and do not know how this process would adhere to the polymer. The polymer could be very gently sanded with a nick brush to give it more tooth and surface area. This may work very well to hold the platinum. You will have to try it out and see – I have not yet had time yet to experiment with various loadings.

Matching gaskets to MEAs

When you do make your own MEAs they may come out thinner or thicker than the .020 silicone gasket material – if that is the case you may have to adjust your gasket material on the hydrogen side for convective cells or both sides for oxygen hydrogen cells to be sure the conductive graphite flow field surface is contacting the carbon cloth well. If your MEAs are a bit on the thin side, McMaster-Carr has a latex in .004, 006, and 008" sizes, that, although not having the temperature range of silicon, will suffice for experimental cells. You can also tighten the fasteners a bit. Latex has an upper maximal operating safe temperature mark of about +180°F. Silicone has an upper limit of around +425°F. At any rate, varying thickness of MEAs can be a problem. When I first put together a few MEAs. I used the .020 silicon to surround the hydrogen side only to find that my MEA was not connecting with the graphite plate very well at all, thus making a low voltage and current reading. Make sure that all your parts are talking to each other.

What to expect from your MEA

As a guideline, one half to one watt output for this size of active area is respectable. On your first try with this amount of tetrachloroplatinate, expect the voltage readings to be good, but the current density to be low. The first attempt at making MEAs should be to get acquainted with the process with as little loading as possible so as not to waste too much platinum. After you are acquainted with the steps in the process and know that its working, you can move on to improve current density.

For testing your MEAs and fuel cells you will need a multimeter which can test both voltage and current. The multimeter should at least be able to read over 2 amps. Look for a multimeter with a 10 to 20 amp range. This gives you the ability to test cells in parallel where more current is produced. To monitor both voltage and current at the same time you will need two multimeters or one ammeter and one voltmeter. You can find most of the information regarding basic testing procedures on the internet in basic electrical and electronic tutorials if you do not have knowledge in this area.

When you first supply the reactants (gases) to a cell of the size we have been working with, they will read anywhere between 1.25 volts to .7 volts with current (amperage) being variable depending on the density of loading. As the voltage drops, the current rises until the cell settles in around the .5 volt range at a higher (determined by density of loading) current under full load conditions.

Testing fuel cells

Testing fuel cells is a whole subject in itself. There are many factors that can affect fuel cell performance. The design of every component in the cell has an effect on the overall output of the cell. Then, there is the question of whether the components are performing their intended function, for instance, are the fasteners screwed tight enough so that the gaskets are keeping gasses from leaking.

Other factors that affect fuel cell performance are operating temperature, under-hydration, and flooding. Membranes need to be hydrated properly to maintain proton conductivity. The membrane can be damaged if the cell becomes too dry. On the other hand, if the membrane floods, the transport of gases is reduced. And, to produce energy most efficiently the best operating temperatures are from 140°F to 175°F.

Gas flow and pressure also affect output. For convection stacks of 3 watts or more, air flow can be an issue and forced air housings are used to help get the oxygen to the cells. Very small low current fans that can run off a small solar cell can also solve that problem. There are other ways to approach this problem, such as with a no-power-needed-design that draws air through a housing, based on the convection process.

To understand all the factors that will affect a fuel cell you need to understand basic physics, basic electricity, and basic chemistry.

Basic test equipment

You can perform all the tests necessary with a bit of basic equipment. For instance, to test the effect of temperature on cell output, a simple temperature monitor such as a thermometer, or thermistor circuit will do. To test pressure, a pressure transducer will suffice, or there is Kynar, a thin film polymer that produces voltage from induced pressure via the piezo-electric effect. Because of its thin profile Kynar can be easily sandwiched in a fuel cell to indicate proper pressure settings for screw adjustments.

To test rate of gas flow, a flow meter is used. Humidity can be monitored with an electric humidity sensor. A handy device to build is an MEA test fixture into which you can easily insert and remove MEAs with a variety of parameters that can be controlled. You can pursue testing and monitoring further in the resource section of this book.

Many of the test devices are suitable to build yourself. If you are new to sensors, pick up a copy of **Sensors** by Forrest Mims, and also a copy of his basic electronics tutorial and other books. It is a good starting point for building and applying sensor technology to your needs. You can also go to the **Sensors Magazine** web site where they have a buyers guide and lots of other good info.

Resources

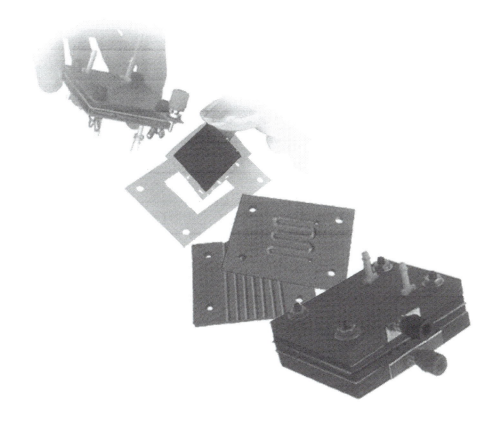

Online Resources

General Fuel Cell Information

US Department of Energy	General information
Fuel Cell Today	General information
Smithsonian Institution	History and general information

Carbon and Graphite Suppliers

Fiberglast	Carbon cloth
Ohio Carbon Blank	Graphite
Fuel Cells Etc	Carbon cloth
Aerocon	Graphite
Fuel Cell Earth	Graphite and carbon cloth

Chemical, Metal, Catalyst Suppliers

Alfa Aesar	Chemical, metals and catalyst suppliers
Sigma-Aldrich	Chemical suppliers
Surepure Chemetals	Platinum metal products
Johnson Matthey	Platinum group metals, catalysts
Fuel Cell Earth	Catalysts

Conversion Tables

OnlineConversion.com	Handy calculator to convert length, distance, weight, speed, volume, area, power, etc., from one unit to another.

Electrical/Electronic Tutorials and Information

Tony Kuphaldt, DC Circuits	Online tutorial

Electrochemistry

Journal of the Electrochemical Society
Index of papers about electrochemical technology

Electrolyzer Suppliers
Fuel Cell Earth — Electrolyzer suppler

Electronic and electrical supplies
Electronic Goldmine — Circuit board and other electronic items
All Electronics — Electronic and electrical items

Electroplating
finishing.com — Tutorial on electroplating
Caswell Plating — Supplier of electroplating equipment and chemicals

Fuel Cells, Component Suppliers, and Information
Fuel Cell Etc — All things related to fuel cells
Golden Energy Fuel Cell — Supplier of MEAs and MEA components
Gore Fuel Cell Technologies — Supplier of MEAs and components
Ion Power — Fuel cell components and MEAs
3M Worldwide — Supplier of MEAs and components
Dupont — Supplier of MEAs and components
Toray — Toray carbon paper information
Fuel Cell Earth — Fuel cell components

Laboratory Supplies
Daigger Lab Equipment & Supplies
Scientific Equipment of Houston
Science Gear
eBay
Amazon

Material Suppliers – Hardware

MSC Direct   General material supplier

McMaster-Carr   General material supplier

Western Enterprises   Gas flow devices, flash arrestors etc.

Materials Information

MatWeb   Database on material properties. Excellent resource to learn about different materials.

Metal Leaf Suppliers

Imai Kinpaku

Wrights of Lymm

Stuart R Stevenson

Platinum Process

Bostick & Sullivan   Platinum process supplies

Surplus

Surplus Ctr. of Nebraska Lots of items for the experimenter

Vacuum Deposition Information

The Bell Jar

Books

The Bell Jar Project Series
by Stephen P. Hansen

Biomimicry
by Janine M. Benyus
William Morrow & Co. 1997
New York

Engineer's Mini-Notebook
by Forrest M. Mims III
Master Publishing Inc. 1996
Lincolnwood IL

Getting Started in Electronics
by Forrest M. Mims III
Master Publishing Inc. 1983
Lincolnwood IL

Platinum and Palladium Printing
by Dick Arentze
Focal Press 2005
Burlington MA

High Voltage Capacitor Design Handbook
by Kenneth R. Scott
Lambda Publishing Group
P.O. Box 1894, Lawrence, KS 66044

High Voltage Experimenter's Handbook
by Jim Lux
(available online)

Homemade Lightning
by R. A. Ford
TAB Books, 1991
Blue Ridge Summit PA

Procedures in Experimental Physics
by John Strong
Prentice Hall (1938)

Sensors
by Forrest M. Mims III
Master Publishing Inc
Lincolnwood IL

Solar Electricity
by Simon Roberts
Prentice Hall, 1991
Hemel Hempstead UK

Build a Solar Hydrogen Fuel Cell System
by Phillip Hurley.
Wheelock Mtn. Publications, 2004
Wheelock VT

Practical Hydrogen Systems: an Experimenter's Guide
by Phillip Hurley
Wheelock Mtn. Publications, 2005
Wheelock VT

K18 - Suppliers for Tools & Materials

Micro-Mark
340 Snyder Avenue
Berkeley Heights, N.J. 07922
800-225-1066

McMaster-Carr Supply Company
473 Ridge Road
Dayton NJ 08810-0317
732-329-3200

MSC Industrial Supply Co.
75 Maxess Road
Melville NY 11747-9415
800-645-7270

Small Parts Inc.
13980 N.W. 58th Court
P.O. Box 4650
Miami Lakes, FL 33014-0650
305-557-7955 or 800-220-4242

Northern Tool and Equipment
P.O. Box 1499
Burnsville, MN 55337-0499
800-533-5545

L78 - Suppliers for Tools & Materials

McMaster-Carr Supply Company
473 Ridge Road
Dayton NJ 08810-0317
732-329-3200

Small Parts Inc.
13980 N.W. 58th Court
P.O. Box 4650
Miami Lakes, FL 33014-0650
305-557-7955 or 800-220-4242

Micro-Mark
340 Snyder Avenue
Berkeley Heights, N.J. 07922
1-800-225-1066

All Electronics
905 S. Vermont Avenue
Log Angeles, CA 90006
213-380-8000

The Electronic Goldmine
P.O. Box 5408
Scottsdale, AZ 85261
800-445-0697

For resources local to you for both fuel cell designs, check out electronics suppliers like Radio Shack; and hardware stores and plumbing supply stores, etc.

Suppliers of MEAs

The fuel cell templates in this book are designed for an MEA with an active area of 1¼" x 1¼" (3.175 cm x 3.175 cm), or 10 square cm. They can accommodate an active area of up to 1⅜" x 1⅜" (3.49 cm x 3.49 cm) or 12.2 square centimeters. The ionomer should extend at least ⅜" beyond all edges of the active area, so the overall dimensions would be at least 1⅝" x 1⅝" for an MEA with a 10 square cm active area.

Ion Power
720 Governor Lea Rd.
New Castle, DE 19720 USA
877-345-9198

Electrochem Inc.
400 W. Cummings Park
Woburn MA 01801 USA
781-938-5300

Fuel Cell Earth
200F Main St. Suite 365
Stoneham MA 02180
617-532-0582

Fuel Cells Etc
979-635-4706

Suppliers of Graphite

Ohio Carbon Blank
38043 Pelton Road
Willoughby OH 44094
800-448-8887
Low cost blanks at approximate size needed, or will cut to exact size.

Aerocon
4170 Golf Drive
San Jose CA 95127
408-926-6507
Variety of shapes: bars, disks, and plates in a variety of sizes.

Fuel Cell Earth
200F Main St. Suite 365
Stoneham MA 02180
617-532-0582.

Material Safety Data Sheets (MSDSs)

You will need to view each MSDS and implement all recommendations and requirements before using, storing, handling and disposing any chemical listed in this book.

It is your responsibility to keep yourself and others safe when working with these substances. These substances are toxic, and some are both toxic and corrosive. You absolutely must wear safety glasses when working with any of these materials. It is good to wear a face shield when working with potassium hydroxide or sulfuric acid.

According to MSDS recommendations, a respirator is required when working with any chemical. Rubber gloves are a must when handling both toxic and corrosive substances listed in this book. Protective clothing per MSDS should be worn. When working with potassium hydroxide and sulfuric acid great care must be taken to protect yourself from these substances.

Some of these substances are toxic – do not eat while working with them. Use a fume hood per recommendation of MSDS. If you do not intend to work with these substances in safe manner DO NOT WORK WITH THEM!!! The author assumes no responsibility for your disregard for safety practices.

Keep all chemicals away from children and pets, and do not allow children or pets into your work area. Do not allow unauthorized persons to enter your work area. Do not allow any smoking or possible source of ignition into your work area when working with hydrogen and oxygen gases.

The MSDSs are available on the internet. Simply search the chemical name and "MSDS".

Potassium tetrachloroplatinate (II)

**Liquid Nafion-perfluorosulfonic acid-PTFE copolymer,
5% w/w solution** (search Alpha Aesar MSDS database for PTFE copolymer)

Ammonium ferric oxalate (ammonium iron (III) oxalate)
(search Alpha Aesar MSDS database)

Potassium hydroxide

Sulfuric acid

Hydrogen peroxide 3%

Hydrogen

Oxygen

Templates

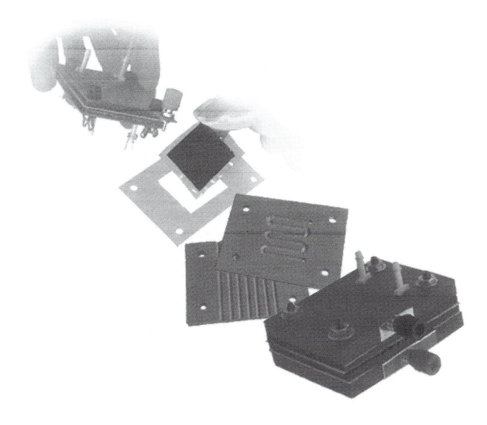

List of Templates

Template A1

K-18 MILLED GRAPHITE FUEL CELL
(SINGLE SLICE CONVECTION)
HYDROGEN SIDE END PLATE

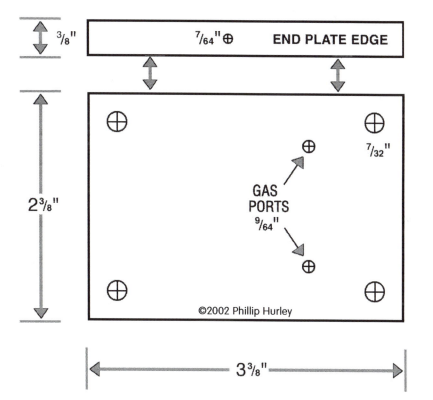

CUT AND DRILL 1 FROM 3/8" THICK FIBERGLASS
OR OTHER SUITABLE MATERIAL.
TEMPLATE SHOWS OUT-FACING SIDE.

179

Template A2

K-18 MILLED GRAPHITE FUEL CELL
(SINGLE SLICE CONVECTION)
ELECTRODES

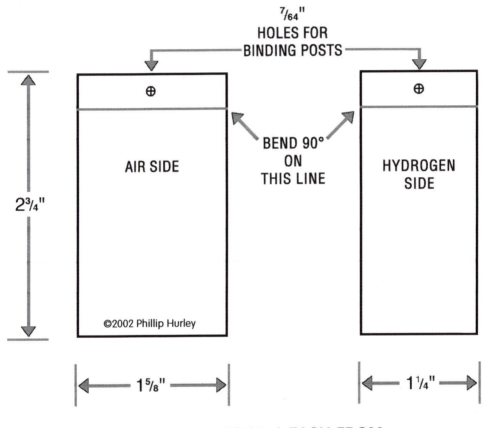

$\frac{7}{64}$"
HOLES FOR
BINDING POSTS

BEND 90°
ON
THIS LINE

AIR SIDE

HYDROGEN
SIDE

2$\frac{3}{4}$"

1$\frac{5}{8}$"

1$\frac{1}{4}$"

©2002 Phillip Hurley

CUT AND DRILL 1 EACH FROM
.010" ALLOY NICKEL OR STAINLESS STEEL

Template A3

K-18 MILLED GRAPHITE FUEL CELL
(SINGLE SLICE CONVECTION)
RUBBER SPACERS

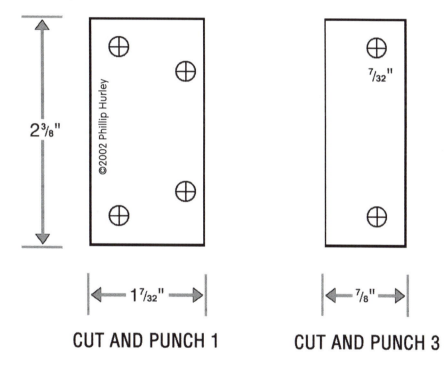

2³⁄₈"

©2002 Phillip Hurley

1⁷⁄₃₂"

⁷⁄₃₂"

⁷⁄₈"

CUT AND PUNCH 1

CUT AND PUNCH 3

.020" SILICONE RUBBER

Template A4

K-18 MILLED GRAPHITE FUEL CELL (SINGLE)
MYLAR SURROUND
AND HYDROGEN SIDE RUBBER GASKET

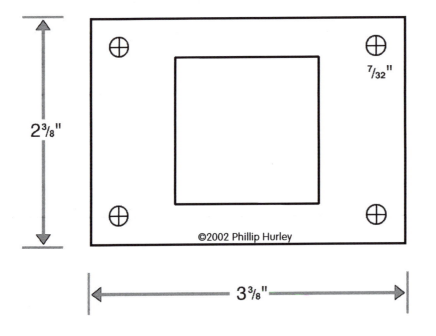

$2\frac{3}{8}"$

$\frac{7}{32}"$

©2002 Phillip Hurley

$3\frac{3}{8}"$

**CUT AND PUNCH 2 FROM MYLAR
AND 1 FROM .020" SILICONE RUBBER**

Template A5

K-18 MILLED GRAPHITE FUEL CELL
(SINGLE SLICE CONVECTION)
GRAPHITE SERPENTINE FLOW PLATE
HYDROGEN SIDE

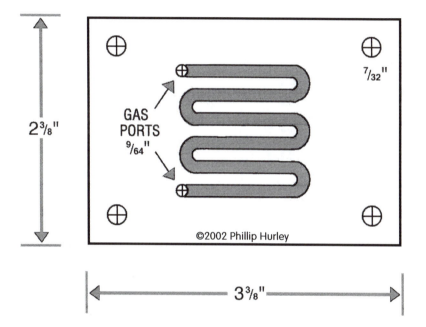

©2002 Phillip Hurley

1 PIECE FROM 3/16" THICK GRAPHITE
DRILL FASTENER HOLES
ROUT GAS GROOVES (SHADED AREA), 1/8" BIT, DEPTH 3/32"

Template A6

K-18 MILLED GRAPHITE FUEL CELL
(SINGLE SLICE CONVECTION)
GRAPHITE VERTICAL FLOW PLATE
AIR SIDE

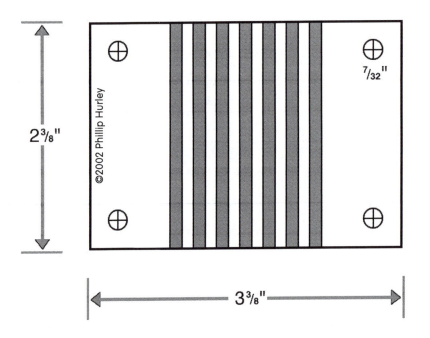

©2002 Phillip Hurley

$7/32"$

$2^{3}/_{8}"$

$3^{3}/_{8}"$

1 PIECE FROM $3/16"$ THICK GRAPHITE
DRILL FASTENER HOLES
ROUT GAS GROOVES (SHADED AREA), $1/8"$ BIT, DEPTH $3/32"$

Template A7

K-18 MILLED GRAPHITE FUEL CELL
(SINGLE SLICE CONVECTION)
AIR SIDE END PLATE

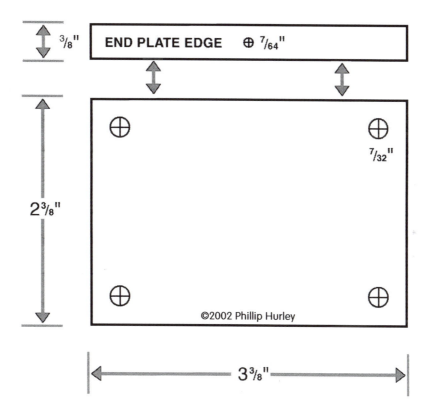

CUT AND DRILL 1 FROM ³/₈" THICK FIBERGLASS
OR OTHER SUITABLE MATERIAL

Template B1

ELECTROLYZER
BOTTOM
(PVC PIPE CAP)

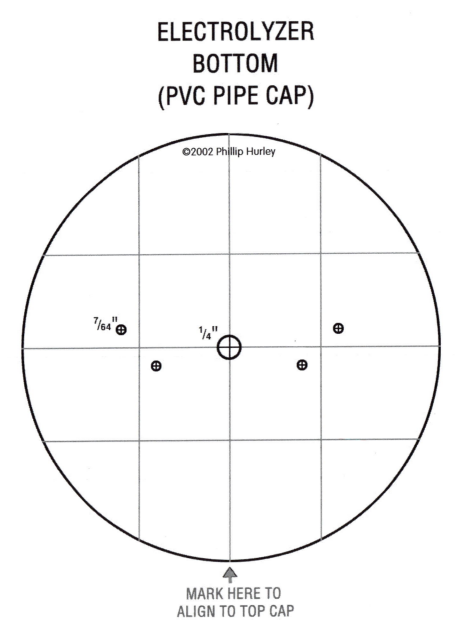

©2002 Phillip Hurley

$7/64$" ⊕ $1/4$" ⊕   ⊕

⊕   ⊕

MARK HERE TO
ALIGN TO TOP CAP

Template B2

ELECTROLYZER
TOP
(PVC PIPE CAP)

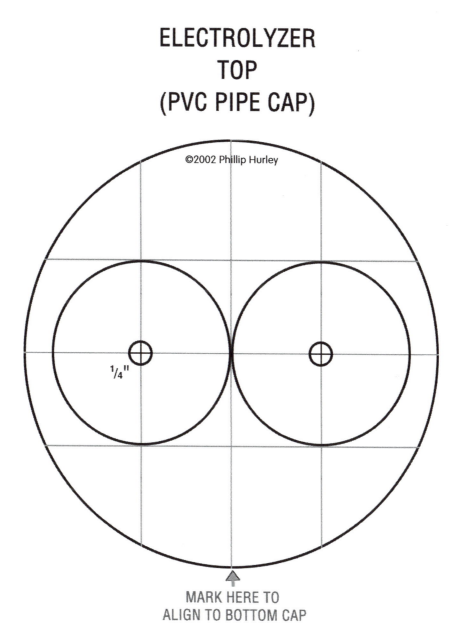

©2002 Phillip Hurley

¼"

MARK HERE TO
ALIGN TO BOTTOM CAP

Template B3

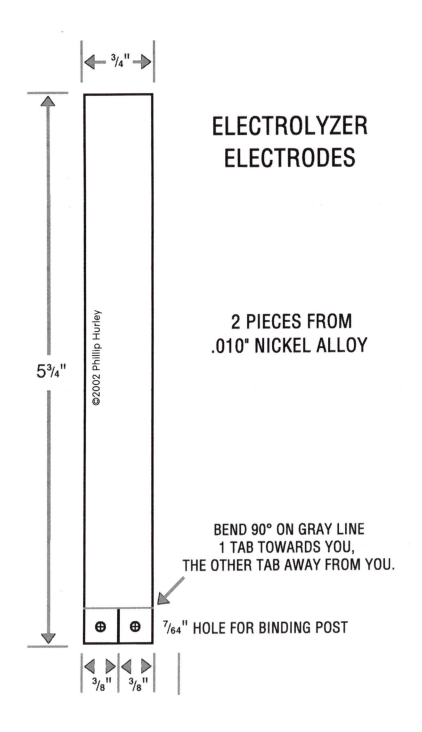

ELECTROLYZER
ELECTRODES

2 PIECES FROM
.010" NICKEL ALLOY

©2002 Phillip Hurley

BEND 90° ON GRAY LINE
1 TAB TOWARDS YOU,
THE OTHER TAB AWAY FROM YOU.

$^7/_{64}$" HOLE FOR BINDING POST

Template C1

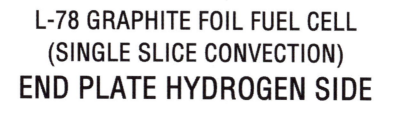

L-78 GRAPHITE FOIL FUEL CELL
(SINGLE SLICE CONVECTION)
END PLATE HYDROGEN SIDE

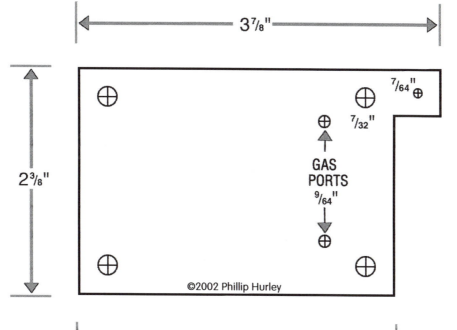

3⁷/₈"

2³/₈"

⁷/₆₄"

⁷/₃₂"

GAS PORTS

⁹/₆₄"

©2002 Phillip Hurley

3³/₈"

1 PIECE CUT FROM SINGLE- OR DOUBLE-SIDED CIRCUITBOARD.
IF DOUBLE-SIDED, REMOVE COPPER SURFACE
FROM ELECTRODE TAB AREA ON OUT-FACING SIDE.
VIEW SHOWS OUT-FACING SIDE.

Template C2

L-78 GRAPHITE FOIL FUEL CELL
(SINGLE SLICE CONVECTION)
GRAPHITE SERPENTINE FLOW PLATE
HYDROGEN SIDE

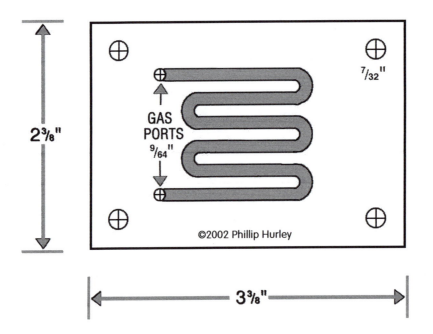

1 PIECE FROM $1/8$" GRAPHITE FOIL
DRILL HOLES
PRESS GROOVES (SHADED AREA) $1/8$" WIDE, DEPTH $1/16$" TO $3/32$"

Template C3

L-78 GRAPHITE FOIL FUEL CELL
(SINGLE SLICE CONVECTION)
MYLAR SURROUND
AND SILICONE RUBBER GASKET

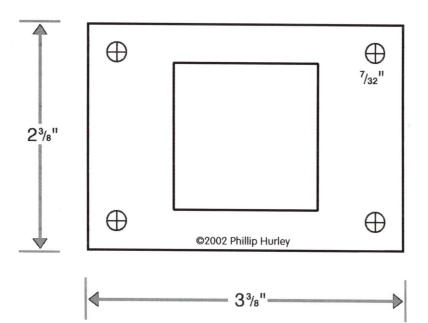

©2002 Phillip Hurley

CUT 2 FROM MYLAR
CUT 1 FROM .020" SILICONE RUBBER

Template C4

L-78 GRAPHITE FOIL FUEL CELL
(SINGLE SLICE CONVECTION)
GRAPHITE VERTICAL FLOW PLATE
AIR SIDE

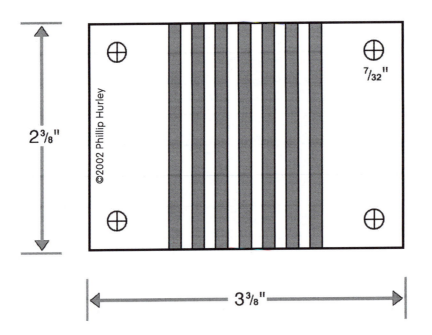

1 PIECE FROM 1/8" GRAPHITE FOIL
DRILL HOLES
PRESS GROOVES (SHADED AREA) 1/8" WIDE, DEPTH 1/16" TO 3/32"

Template C5

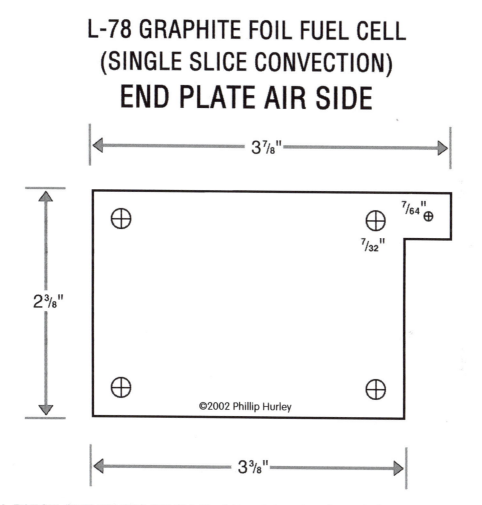

L-78 GRAPHITE FOIL FUEL CELL
(SINGLE SLICE CONVECTION)
END PLATE AIR SIDE

$3\frac{7}{8}$"

$2\frac{3}{8}$"

$\frac{7}{64}$"

$\frac{7}{32}$"

©2002 Phillip Hurley

$3\frac{3}{8}$"

1 PIECE CUT FROM SINGLE- OR DOUBLE-SIDED CIRCUITBOARD.
IF DOUBLE-SIDED, REMOVE COPPER SURFACE
FROM ELECTRODE TAB AREA ON OUT-FACING SIDE.
VIEW IS OF OUT-FACING SIDE.

Template D1

K-18 MILLED GRAPHITE FUEL CELL
(SINGLE SLICE OXYGEN/HYDROGEN)
END PLATES

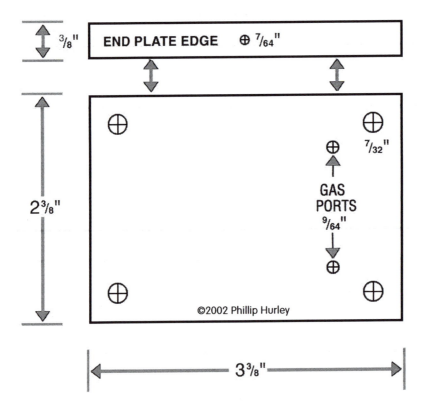

CUT AND DRILL 2 FROM 3/8" THICK FIBERGLASS
OR OTHER SUITABLE MATERIAL.
TEMPLATE SHOWS OUT-FACING SIDE

Template D2

K-18 MILLED GRAPHITE FUEL CELL
(SINGLE SLICE OXYGEN/HYDROGEN)
ELECTRODES

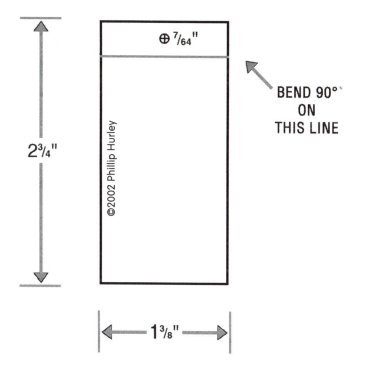

CUT AND DRILL 2 FROM
.010" ALLOY NICKEL OR STAINLESS STEEL

Template D3

K-18 MILLED GRAPHITE FUEL CELL
(SINGLE SLICE OXYGEN/HYDROGEN)
RUBBER SPACERS

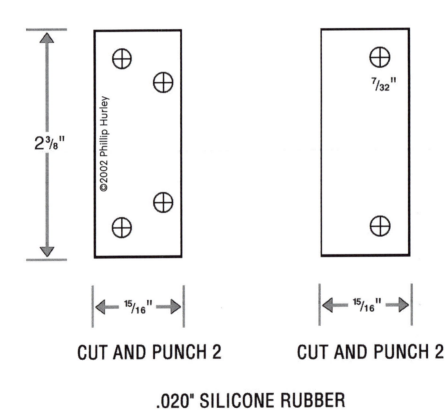

©2002 Phillip Hurley

2 3/8"

15/16"

CUT AND PUNCH 2

7/32"

15/16"

CUT AND PUNCH 2

.020" SILICONE RUBBER

Template D4

K-18 MILLED GRAPHITE FUEL CELL
(SINGLE SLICE OXYGEN/HYDROGEN)
GRAPHITE SERPENTINE FLOW PLATES

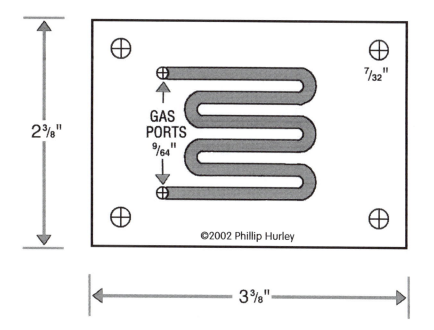

2 PIECES FROM 3/16" THICK GRAPHITE
DRILL FASTENER HOLES
ROUT GAS GROOVES (SHADED AREA), 1/8" BIT, DEPTH 3/32"

Template D5

K-18 MILLED GRAPHITE FUEL CELL
(SINGLE SLICE OXYGEN/HYDROGEN)
MYLAR SURROUND
AND HYDROGEN SIDE RUBBER GASKET

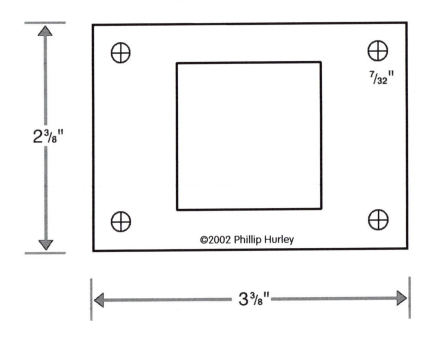

©2002 Phillip Hurley

CUT AND PUNCH 2 FROM MYLAR
AND 2 FROM .020" SILICONE RUBBER

Template E1

L-78 GRAPHITE FOIL FUEL CELL
(SINGLE SLICE OXYGEN/HYDROGEN)
END PLATES

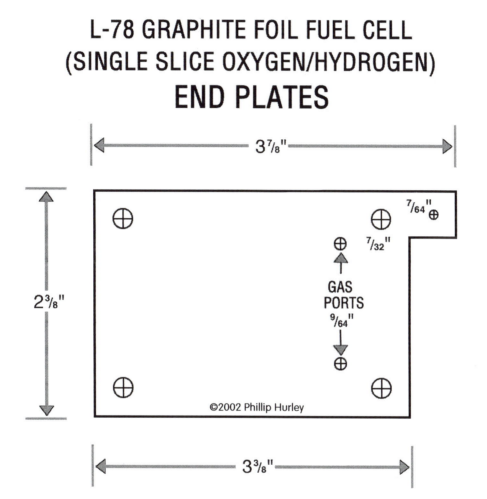

2 PIECES CUT FROM SINGLE- OR DOUBLE-SIDED CIRCUITBOARD.
IF DOUBLE-SIDED, REMOVE COPPER SURFACE
FROM ELECTRODE TAB AREA ON OUT-FACING SIDE.
TEMPLATE SHOWS OUT-FACING SIDE.

Template E2

L-78 GRAPHITE FOIL FUEL CELL
(SINGLE SLICE OXYGEN/HYDROGEN)
GRAPHITE SERPENTINE FLOW PLATE

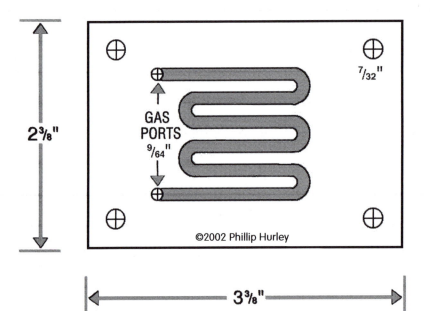

2 PIECES FROM $1/8$" GRAPHITE FOIL
DRILL HOLES
PRESS GROOVES (SHADED AREA) $1/8$" WIDE, DEPTH $1/16$" TO $3/32$"

Template E3

L-78 GRAPHITE FOIL FUEL CELL
(SINGLE SLICE OXYGEN/HYDROGEN)
MYLAR SURROUND
AND SILICONE RUBBER GASKET

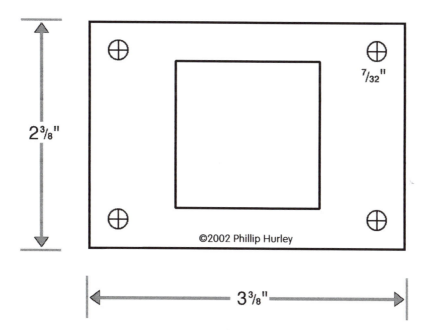

©2002 Phillip Hurley

CUT 2 FROM MYLAR
CUT 2 FROM .020" SILICONE RUBBER

Template F1

K-18 MILLED GRAPHITE FUEL CELL
(CONVECTION STACK)
ENDPLATE, HYDROGEN SIDE

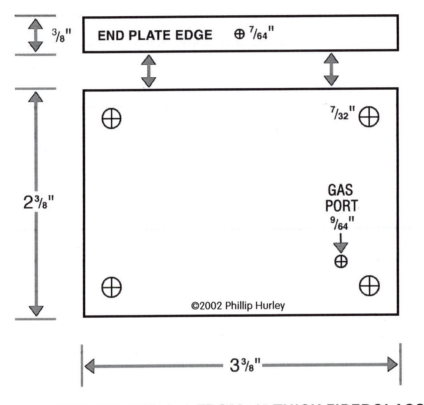

CUT AND DRILL 1 FROM 3/8" THICK FIBERGLASS
OR OTHER SUITABLE MATERIAL.
TEMPLATE SHOWS OUT-FACING SIDE.

Template F2

K-18 MILLED GRAPHITE FUEL CELL
(CONVECTION STACK)
RUBBER SPACERS

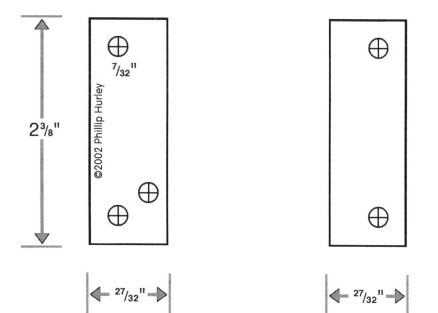

CUT AND PUNCH 2 OF EACH
.020" SILICONE RUBBER

Template F3

K-18 MILLED GRAPHITE FUEL CELL
(CONVECTION STACK)
ELECTRODES

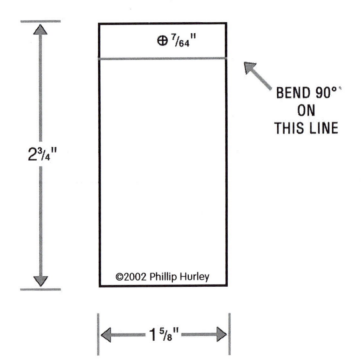

⊕ $^7/_{64}$"

BEND 90°
ON
THIS LINE

$2^3/_4$"

©2002 Phillip Hurley

←— $1^5/_8$" —→

CUT AND DRILL 2 FROM
.010" ALLOY NICKEL OR STAINLESS STEEL

Template F4

K-18 MILLED GRAPHITE FUEL CELL
(CONVECTION STACK)
GRAPHITE SERPENTINE FLOW PLATE
(FIRST PLATE, HYDROGEN SIDE)

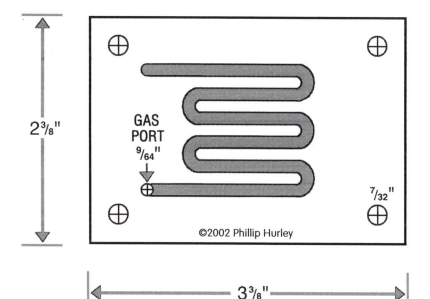

1 PIECE MILLED ON ONE SIDE ONLY
FROM 3/16" THICK GRAPHITE
DRILL FASTENER HOLES
ROUT GAS GROOVES (SHADED AREA), 1/8" BIT, DEPTH 3/32"

Template F5

K-18 MILLED GRAPHITE FUEL CELL
(CONVECTION STACK)
MYLAR SURROUND & RUBBER GASKET

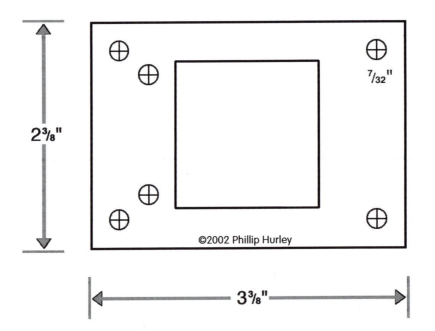

©2002 Phillip Hurley

CUT AND PUNCH 2 FROM MYLAR
AND 2 FROM .020" SILICONE RUBBER
FOR EACH PEMBRANE

Template F6

K-18 MILLED GRAPHITE FUEL CELL
(CONVECTION STACK)
BIPOLAR PLATE, HYDROGEN SIDE

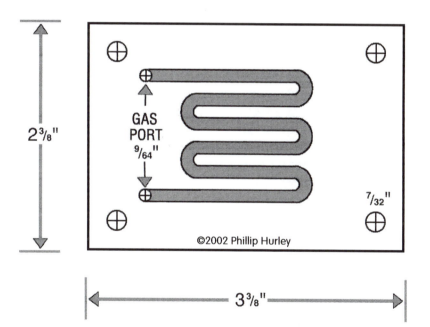

©2002 Phillip Hurley

$^{5}/_{16}$" THICK GRAPHITE
DRILL FASTENER HOLES
ROUT GAS GROOVES (SHADED AREA), $^{1}/_{8}$" BIT, DEPTH $^{3}/_{32}$"

Template F7

K-18 MILLED GRAPHITE FUEL CELL
(CONVECTION STACK)
BIPOLAR PLATE, HYDROGEN SIDE

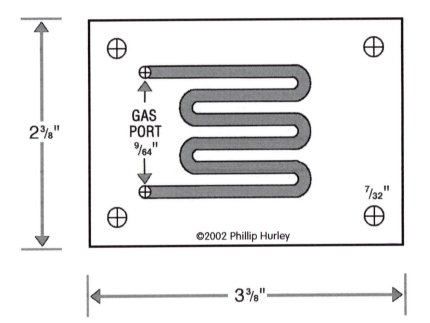

$5/16$" THICK GRAPHITE
DRILL FASTENER HOLES
ROUT GAS GROOVES (SHADED AREA), $1/8$" BIT, DEPTH $3/32$"

Template F8

K-18 MILLED GRAPHITE FUEL CELL
(CONVECTION STACK)
GRAPHITE VERTICAL FLOW PLATE
(LAST PLATE, AIR SIDE)

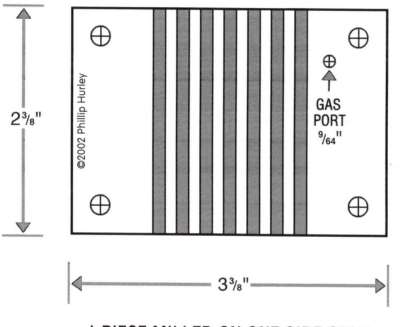

1 PIECE MILLED ON ONE SIDE ONLY
$3/16$" THICK GRAPHITE
DRILL FASTENER HOLES
ROUT GAS GROOVES (SHADED AREA), $1/8$" BIT, DEPTH $3/32$"

Template F9

K-18 MILLED GRAPHITE FUEL CELL
(CONVECTION STACK)
ENDPLATE, AIR SIDE

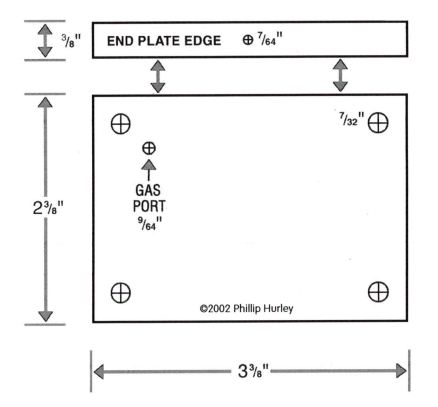

CUT AND DRILL 1 FROM ³/₈" THICK FIBERGLASS
OR OTHER SUITABLE MATERIAL.
TEMPLATE SHOWS OUT-FACING SIDE.

Template G1

K-18 MILLED GRAPHITE FUEL CELL
(OXYGEN/HYDROGEN STACK)
ENDPLATES

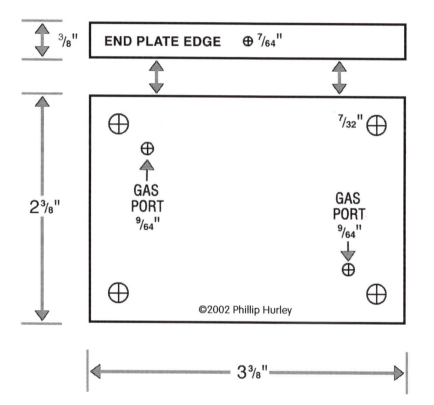

CUT AND DRILL 2 FROM 3/8" THICK FIBERGLASS
OR OTHER SUITABLE MATERIAL.
TEMPLATE SHOWS OUT-FACING SIDE.

Template G2

K-18 MILLED GRAPHITE FUEL CELL
(OXYGEN/HYDROGEN STACK)
ELECTRODES

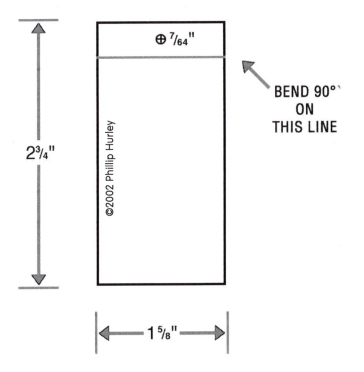

BEND 90°
ON
THIS LINE

⊕ 7/64"

2 3/4"

1 5/8"

©2002 Phillip Hurley

CUT AND DRILL 2 FROM
.010" ALLOY NICKEL OR STAINLESS STEEL

Template G3

K-18 MILLED GRAPHITE FUEL CELL
(OXYGEN/HYDROGEN STACK)
GRAPHITE SERPENTINE FLOW PLATE
(FIRST AND LAST PLATE)

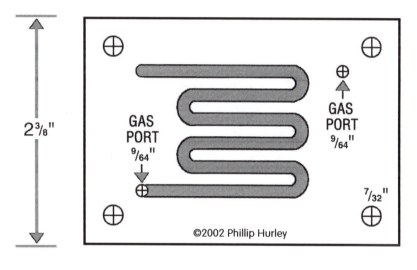

2 PIECES MILLED ON ONE SIDE ONLY
FROM 3/16" THICK GRAPHITE
DRILL FASTENER HOLES
ROUT GAS GROOVES (SHADED AREA), 1/8" BIT, DEPTH 3/32"

©2002 Phillip Hurley

Template G4

K-18 MILLED GRAPHITE FUEL CELL
(OXYGEN/HYDROGEN STACK)
GRAPHITE SERPENTINE FLOW PLATE
(FIRST AND LAST PLATE)

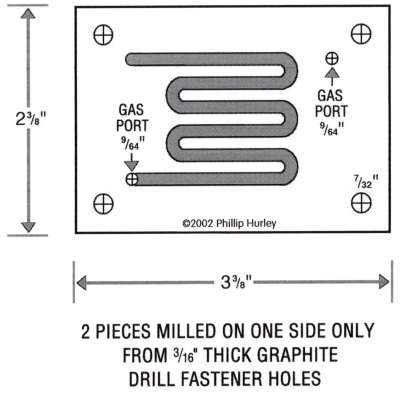

2 PIECES MILLED ON ONE SIDE ONLY
FROM 3/16" THICK GRAPHITE
DRILL FASTENER HOLES
ROUT GAS GROOVES (SHADED AREA), 1/8" BIT, DEPTH 3/32"

Template G5

K-18 MILLED GRAPHITE FUEL CELL
(OXYGEN/HYDROGEN STACK)
MYLAR SURROUND & RUBBER GASKET

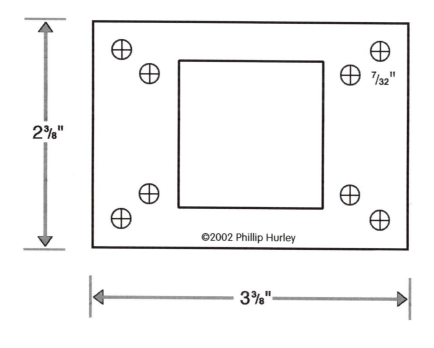

©2002 Phillip Hurley

CUT AND PUNCH 2 FROM MYLAR
AND 2 FROM .020" SILICONE RUBBER
FOR EACH PEMBRANE

Template G6

K-18 MILLED GRAPHITE FUEL CELL
(OXYGEN/HYDROGEN STACK)
GRAPHITE SERPENTINE FLOW PLATES
(BIPOLAR PLATES)

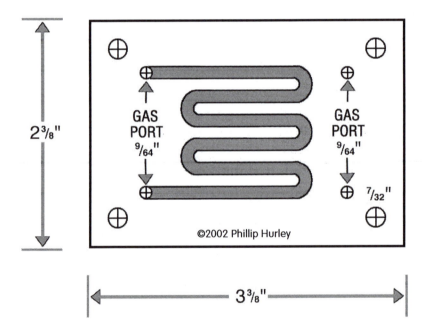

GAS PORT $9/64$"

GAS PORT $9/64$"

$7/32$"

$2^3/8$"

$3^3/8$"

©2002 Phillip Hurley

$5/16$" THICK GRAPHITE
DRILL FASTENER HOLES
ROUT GAS GROOVES (SHADED AREA), $1/8$" BIT, DEPTH $3/32$"

Template H1

SURROUND GASKET
ALTERNATIVE TO
GRAPHITE SERPENTINE FLOW PLATES

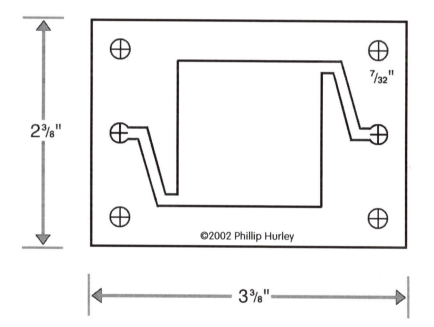

CUT AND PUNCH FROM .020" SILICONE RUBBER

Build Your Own Solar Panel

by Phillip Hurley

Whether you're trying to get off the grid, or you just like to experiment, *Build Your Own Solar Panel* has all the information you need to build your own photovoltaic panel to generate electricity from the sun. The new revised and expanded edition has easy-to-follow directions, and over 150 detailed photos and illustrations. Materials and tools lists, and links to suppliers of PV cells are included. Every-day tools are all that you need to complete these projects.

Build Your Own Solar Panel will show you how to:

- Design and build PV panels
- Customize panel output
- Make tab and bus ribbon
- Solder cell connections
- Wire a photovoltaic panel
- Purchase solar cells
- Test and rate PV cells
- Repair damaged solar cells
- Work with broken cells
- Encapsulate solar cells

Available in print from Amazon.com and for download in full color PDF ebook format at

www.buildasolarpanel.com

Download a free sample of Build Your Own Solar Panel in full color PDF at www.buildasolarpanel.com.

Solar II:

How to Design, Build and Set Up Photovoltaic Components and Solar Electric Systems

by Phillip Hurley

Now that you've built your solar panels, how do you set up a PV system and plug in? In the e-book Solar II, Phillip Hurley, author of *Build Your Own Solar Panel*, will show you how to:

- ◆ Plan and size your solar electric system
- ◆ Build racks and charge controllers
- ◆ Mount and orient PV panels
- ◆ Wire solar panel arrays
- ◆ Make a ventilated battery box
- ◆ Wire battery arrays for solar panels
- ◆ Install an inverter
- ◆ Maintain solar batteries for optimum life and performance
- ◆ Make your own combiner box, bus bars and DC service box

Solar II includes over 150 photos and illustrations, and a daily power usage calculator. Published April 2007.

Available in print from Amazon.com and for download in full color PDF ebook format at :

www.buildasolarpanel.com

*Download a free sample of Solar II in full color PDF
at www.buildasolarpanel.com.*

The Battery Builder's Guide
by Phillip Hurley

The Battery Builder's Guide is a practical hands-on text that will show you how to make your own rechargeable flooded lead acid batteries. Learn how to recycle parts and materials, how to fabricate battery components and where to purchase the parts, materials and tools you need to build or rebuild batteries. The text covers construction of batteries with Plante (pure lead) and Faure (pasted lead) plates.

Topics include:

- Recycling old lead acid batteries
- Molding battery parts
- Design formulas and tables
- Lead burning
- Techniques and tools for battery building
- Building plate burning racks
- Pasting and forming plates
- Types of batteries such as SLA and deep cycle, and their characteristics and uses
- And more... all illustrated with extensive step-by-step photos

Flooded lead acid batteries are used for stationary applications such as solar and wind powered electrical systems, and for mobile applications. If you need custom batteries of a specific size or output, wish to experiment with building batteries, or want to lower your costs by using recycled components and materials, *The Battery Builder's Guide* has the information you need.

Available in print from Amazon.com and for download in full color PDF ebook format at

www.batterybuildersguide.com

Download a free sample of The Battery Builders Guide in full color PDF at www.batterybuildersguide.com.

Build A Solar Hydrogen Fuel Cell System

by Phillip Hurley

Learn how to construct and operate the components of a solar hydrogen fuel cell system: the fuel cell stack, the electrolyzer to generate hydrogen fuel, simple hydrogen storage, and solar panels designed specifically to run electrolyzers for hydrogen production. Complete, clear, illustrated instructions to build all the major components make it easy for the non-engineer to understand and work with this important new technology.

Featured are the author's innovative and practical designs for efficient solar powered hydrogen production including:

◆ ESPMs (Electrolyzer Specific Photovoltaic Modules) – 40 watt solar panels designed specifically to run electrolyzers efficiently;

◆ a 40-80 watt electrolyzer for intermittant power from renewable energy sources such as solar and wind;

◆ and, a 6-12 watt planar hydrogen fuel cell stack to generate electricity.

Any of these components can be ganged or racked, or scaled up in size for higher output. You'll also learn how to set up an entire gas processing system, and where to find parts and materials – everything you need for an experimental stationary unit that will give you a solid base for building and operating systems for larger power needs. There are even schematics for adapting conventional solar panels (BSPMs – Battery Specific Photovoltaic Modules) for efficient hydrogen production, and setting up hybrid (battery and fuel cell) PV systems.

Available in print from Amazon.com and for download in full color PDF ebook format at

www.solarh.com

Titles from

Wheelock Mountain Publications:

Other titles by Phillip Hurley:

Solar II

Build Your Own Solar Panel

Build a Solar Hydrogen Fuel Cell System

Practical Hydrogen Systems

Build Your Own Fuel Cells

The Battery Builder's Guide

Solar Supercapacitor Applications

and also:

Solar Hydrogen Chronicles *edited by Walt Pyle*

Tesla: the Lost Inventions *by George Trinkaus*

Tesla Coil *by George Trinkaus*

Radio Tesla *by George Trinkaus*

Son of Tesla Coil *by George Trinkaus*

www.buildasolarpanel.com

Wheelock Mountain Publications
is an imprint of

Good Idea Creative Services
324 Minister Hill Road
Wheelock VT 05851
USA

80678091R00130

Made in the USA
Middletown, DE
16 July 2018